PRAISE FOR *THE WILDCRAFTING BREWER*

"Pascal Baudar takes wild fermentation to the next level with wild plants, wild yeasts, and wild bacteria. His methods are effective, and his creativity is infectious. With gorgeous photos and clear technical details, this book will be a source of great inspiration."

— Sandor Ellix Katz, author of *Wild Fermentation* and *The Art of Fermentation*

"Owning one of Pascal Baudar's books is like possessing a key to the foraging kingdom—a key that opens the door to his unique approach to working with wild ingredients. Over the years, Baudar has developed an original culinary language of the land, working and exploring every element of terroir from salt and stones to insect sugars and plants. His methods are rigorously researched, and his piercing creativity and spirit of enquiry gives them life. The foraging world owes a great deal to Baudar's original research and generous spirit of sharing."

— Marie Viljoen, urban forager and author of *66 Square Feet*

"Pascal Baudar has elevated the concept of terroir—that intricate symbiosis of *Homo imbibens*, native biota, microorganisms, and landscape—into the realm of extreme beverages, both fermented and unfermented. His book brings to life the innovative quest of the Palaeolithic shaman/healer/brewer."

— Patrick E. McGovern, archaeologist and author of *Ancient Brews* and *Uncorking the Past*

"Pascal's new book offers a wonderfully tangential and unrestrained approach to the world of home brewing by mixing foraging with wild booze alchemy and encouraging the reader to experiment, explore, and play. Whilst managing to make both topics fun and accessible, a comprehensive introduction leads to some delightfully simple recipes and plenty of great ideas, all complemented by beautiful photography."

— John Rensten, founder of Forage London and author of *The Edible City*

"Pascal Baudar eliminates the boundaries set by modern homebrewing 'rules' and encourages brewers to go wild with creativity. Not only does he present simple guidelines for making truly unique brews that blur the lines between wine, beer, mead, and soda, but he provides readers with knowledge on how to brew with ingredients they likely already have access to in their kitchens, gardens, yards, and wildlands."

—Jereme Zimmerman, author of *Make Mead Like a Viking*

"I wish I'd had *The Wildcrafting Brewer* years ago when I became obsessed with plant-forward, foraged, alcohol fermentation. Though often overlooked in contemporary brewing, Pascal Baudar focuses on the basics of foraging and alternative sugar sources. At once looking to indigenous history as well as reviving, or creating, innovative techniques, Pascal encourages people to get their hands dirty, fermenting with what is around them rather than worrying first about fancy tools."

—Pete Halupka, Harvest Roots Ferments

"I admire the foraging practice Pascal Baudar shares in *The Wildcrafting Brewer*—hiking into nature and gathering what's prevalent to create a fermented mélange that carries the terroir of those moments he spent forest bathing, beachcombing, or urban scavenging. The season, the scents, the scenery—all imprinted into his bubbly brews. It inspires me to cleanse my aura with a sage cider cocktail and offer a libation of *Achillea* and *Artemisia* to the folks, like Baudar, who keep the spirited plant traditions going. This is a great reference book for both the herbalist's kitchen and the botanist's bar."

—Jessyloo Rodrigues, herbalist and cocktail farmer

The Wildcrafting Brewer

Also by Pascal Baudar

The New Wildcrafted Cuisine

The Wildcrafting Brewer

Creating Unique Drinks and Boozy Concoctions
from Nature's Ingredients

PASCAL BAUDAR

Chelsea Green Publishing
White River Junction, Vermont

Project Manager: Alexander Bullett
Project Editor: Benjamin Watson
Copy Editor: Laura Jorstad
Proofreader: Paula Brisco
Indexer: Shana Milkie
Designer: Melissa Jacobson

Printed in the United States of America.
First printing February 2018.
10 9 8 7 6   23 24 25 26 27

Our Commitment to Green Publishing

Chelsea Green sees publishing as a tool for cultural change and ecological stewardship. We strive to align our book manufacturing practices with our editorial mission and to reduce the impact of our business enterprise in the environment. We print our books using vegetable-based inks whenever possible. This book may cost slightly more because it was printed on paper from responsibly managed forests, and we hope you'll agree that it's worth it. *The Wildcrafting Brewer* was printed on paper supplied by Versa that is certified by the Forest Stewardship Council.®

Library of Congress Cataloging-in-Publication Data
Names: Baudar, Pascal, 1961– author.
Title: The wildcrafting brewer : creating unique drinks and boozy concoctions
    from nature's ingredients / Pascal Baudar.
Description: White River Junction, Vermont : Chelsea Green Publishing, [2018]
    | Includes bibliographical references and index.
Identifiers: LCCN 2017040066| ISBN 9781603587181 (pbk.) | ISBN 9781603587198 (ebook)
Subjects: LCSH: Brewing. | Liquors. | Ale. | Beer. | LCGFT: Cookbooks.
Classification: LCC TP577 .B355 2018 | DDC 663.31—dc23
LC record available at https://lccn.loc.gov/2017040066

Chelsea Green Publishing
White River Junction, Vermont, USA
London, UK
www.chelseagreen.com

To my dad, Henri Baudar
Thank you for all your help, I miss you terribly . . .

CONTENTS

Sugar Beers, Herbal Meads, Odd Sodas, and Funky Wines

This book is about brewing in general—not just making beer but having fun fermenting all kinds of delicious concoctions with whatever we find on offer from nature, be that the wilderness or our backyards or gardens. We're talking odd, wild, and primitive beers, sodas, herbal meads, inebriating (or not) infusions, and many other types of fermented drinks that are sometimes hard to classify. It's about exploring boozy possibilities in creative ways, the way humans have for a very long time. What's not to like about that?

My big discovery when working on this book was the fact that brewing is really a continuous, linear activity. We like to chop up this creative line into small, discrete segments and impose etiquettes on them: *That's a beer*, *that's a wine*, or *that's a soda*. But the truth is that humans, since the dawn of time, have been brewing boozy concoctions that often transcend regular labels. You'll find all kinds of interesting drinks that are really a blend between beers and wines, or sodas and beers. And it's all good: Brewing should be about creativity, flavors, and in some cases medicinal applications. Like many enjoyable activities, it's a lot less fun when you're told what you can or cannot do.

It's an interesting statement for an author to make, but I would like you to look at this book not in terms of precise recipes you can make at home but more as a book of concepts and ideas that will enable you to explore and create with your own local ingredients. The fun is really in dreaming up and brewing your own delicious drinks, so look at the techniques described here and see what you can come up with!

If there is an overall message I want to convey, it is that brewing is fun, adventurous, and extremely rewarding. Don't be afraid to experiment: You might make a few mistakes here and there, mostly in the beginning, but those will be dwarfed by the countless yummy drinks you'll create and be successful at making. Hopefully the basic procedures in this book will help, and I can't wait to hear about the delicious beverages you'll make with the plants surrounding you.

If you already have experience in making beers, you may find some of the methods described here a bit primitive and unusual: the use of molasses and other sources of sugar (including insect honeydew) instead of malt; wild yeast extraction from local fruits, flowers, and plants; unusual bitter wild plants instead of regular hops; and the overall lack of grains. I hope this book will nevertheless inspire you to explore your own terroir and possibly integrate some local flavors into your brews.

The most important message from this book is really: Explore, have fun, and create! That's the stuff life should be made of.

On Picking Wild Plants, or Foraging

Like many of our human activities, foraging can be done for good or evil: It can help the environment or intensify sustainability issues. Over the years I've learned to streamline my activities so as to minimize my impact on nature. It's been a learning curve with trials and errors, but these days I actually think foraging can be done such that you help your local environment, by removing non-native plants (around 90 percent of what I pick) and harvesting sustainably or growing the native plants you need. As far as I remember, at this point I've pretty much planted all the native berries and plants I use in this book in much larger quantities than I'll ever use, mostly on private lands owned by friends.

You don't need to be a fanatic tree-hugger to see that our planet faces several real problems such as pollution, climate change (naturally occurring or not), human expansion, loss of natural habitat, extinctions, and much more. At this point in our evolution, we absolutely need to be part of the solution, and this even applies to the simple activity of harvesting wild plants. We must make sure that our activity of picking wild plants for food, drinks, or medicine is done carefully with environmental health and integrity in mind.

Picking plants and berries for food or making drinks can connect us back to nature. It is a sacred link that, as a species, we all share. We are here because our ancestors had a very intimate relationship with nature, knew which plants to use for food or medicine, and, in many instances, knew how to sustainably interact with their wild environment. No matter where we live, it's part of our cultural DNA.

I don't think the impulse of trying to save nature by protecting it at all costs with a "look don't touch" mentality will work. Growing up in

Belgium, my love for nature came through a deep interaction with my wild surroundings. If you truly love something, you will take care of it and make sure it's still there for generations to come.

When I was a kid, raising animals, growing food in our garden, and picking wild berries, nuts, and plants wasn't considered weird or special—it was a normal part of life. The knowledge was kept by elders, who would pass it on to the next generation. In many modernized countries this cycle of transferring knowledge has been lost. Very valuable and nutritious food plants such as dandelion, mallow, and others are looked upon as "weeds," and TV commercials gladly promote the use of toxic chemicals to destroy them. The people I've seen trashing the wilderness are the product of our current society: If you don't know or understand the value of something, you simply won't care for it.

So do it the right way! Respect the environment, learn which plants are rare or illegal to pick, don't forage plants in protected areas (natural preserves and the like), work with native plant nurseries, and educate yourself on how to grow native plants and remove non-natives.

If you take from nature, work with her and make sure you always plant more than you'll ever harvest so that future generations will still have the same creative opportunities that you do—or more.

CHAPTER 1

Rediscovering the Past

In my last book, *The New Wildcrafted Cuisine*, I started exploring the concept of creating alcoholic beverages using local plants, sugar sources, and even yeasts and bacteria to represent, through taste, the various environments or regions we live in.

For example, if you live in Vermont, Maine, or New Hampshire, you can create delicious fermented beverages or "beers" representing a specific forest, mountain, or territory by using what's available in that location: maple or birch syrup, wild sassafras roots, spruce, birch bark, white pine, yarrow, turkey tail or chaga mushrooms, local yeast from flowers, wild berries, or barks, and so on. You could call it *hyperlocal brewing*. I've found from my own actual research and experience that the number of possible ingredients you can use is mind boggling.

Based on the amount of emails and feedback I received, the brewing section of that first book ended up being quite popular, and probably for good reason. Many people (myself included) find it thrilling to be able to explore a location and create drinks such as beers, wines, or sodas that are a true representation of the site.

From a sustainability perspective, once you find local ingredients that you can use, you can also replant them and thus contribute to their long-term viability. Some of the people who have attended my classes and workshops are now growing native plants in their gardens or on private properties for culinary uses. It's another way you can help the local environment and support its unique native flora and fauna.

In the last couple of years, I've done much more experimentation with various types of brewing and wild yeast extractions, and have traveled to other areas of the country and Europe. Today I'm still as excited as I was when I started years ago to create interesting fermented concoctions with what each

Primitive brewing in a clay pot—open fermentation.
Pottery made by Melissa Brown Bidermann.

specific region has to offer. While teaching food preservation and fermentation at an environmental college in Vermont, after a few days of education about the local flora from fellow plant experts, we were able to create a primitive beer using plants, bark, and berries from their own private forest, extract local wild yeast from flowers, and use their homemade maple syrup (from the same forest) for fermentation. It was so good, I'm still dreaming about it.

In this book, my interest is to share with you not only the techniques I use but also the philosophy and approach behind them. Wildcrafted brewing is really a work in progress—and if you think about it, fermentation has been a work in progress since the dawn of humanity. To this day, thousands of years after the first primitive brews were made, we still obsess about creating the perfect beer or wine.

Honestly, when writing *The New Wildcrafted Cuisine*, I thought that my approach to cooking and creating alcoholic beverages was probably a bit too eccentric or unusual, but it ended up being well received. Since I was a kid, I've taken pleasure in researching, experimenting with, and mixing things. Fifty-five years later it seems that nothing much has changed. I find it thrilling to research new flavors and create interesting condiments for people to experience. And of course I use the same approach of mixing methods, ingredients, and techniques with beverages—especially fermented ones.

In fact, I even mix the definitions of *brewing*. Take a look at the two definitions of the word:

BREWING:
(1) to prepare (as beer or ale) by steeping, boiling, and fermentation or by infusion and fermentation.
(2) to make (a beverage) by boiling, steeping, or mixing various ingredients: brew tea.

Most of my alcoholic beverages—sodas, wines, and what I somewhat loosely call beers—are closer to fermented infusions of a sort: mixes of plants, sugar sources, fruits, and sometimes malted grains designed to achieve interesting flavors. As a wild food instructor and culinary researcher, I tend to concentrate on what I call the true flavors of a local "terroir," which, from my perspective, means using what can be found in the original landscape as opposed to nonlocal or recently imported plants and ingredients (which would therefore include such things as hops).

For example, when I was visiting Vermont one June, making a local forest beer meant brewing with blue spruce branches and needles, wild sassafras roots, chaga mushrooms, white pine needles, maple syrup, local yeast from dandelion flowers, yellow birch bark and branches, and so on. If

I had showed up there in September, due to the different flora in late summer, I probably would have ended up brewing a completely different type of beer, which would have included local berries, for example. You can make many types of "forest beer" from the same forest, and each batch will have different flavors based on the season.

By the way, and just to make it clear, there is absolutely nothing wrong with brewing regular beers; in fact, as a good Belgian, I enjoy them very much and love to brew some as well. We're just dealing with a different approach, very much based on nature and the bounty she provides.

To give a bit of background information and so you understand my approach a bit more, I became interested in the subject of brewing around ten years ago. However, at the time I wasn't especially interested in making sodas, beers, or wines; I was in the process of researching all the various methods of food preservation that could be applied to local wild plants, and I saw brewing as another way to preserve what I was collecting. Pretty much in the same category as pickling, canning, dehydrating, or fermenting.

In my obsession to find culinary uses and preservation methods for my foraged goodies, though, making fermented drinks based on what I was able to collect quickly rose pretty high on my list. If you know how to make interesting and delicious fermented drinks, you can create countless other ingredients and condiments such as sauces, vinegars, and pickles, or even use them for cooking.

Given my upbringing in Belgium, and at the same time being so close to France (the French customs office was literally on the other side of the street), I was familiar with cooking with wine or beers; it was a very traditional way to add interesting and local flavors to food. One of my favorite dishes that my father used to make was a traditional French/Belgian recipe for Easter: roasted leg of lamb cooked in the oven with beer and Dijon mustard. I don't recall the beer he used exactly, probably a type of fruity lambic, but the dish was so good that leftovers would never last very long in the fridge. My trusted companion, a Jack Russell terrier, would make a point of waking me up in the middle of the night, and together we would make quick work of the yummy leftover lamb, leaving no evidence behind after the raid but a clean bone in my dog's bed.

My approach was to study brewing from a forager's perspective. It was an unusual point of view with a strong emphasis on utilizing what I was able to find.

You may think there aren't that many things out there to brew with, but after many years of research and experimenting, I've found well over 150 possible ingredients that can be used in the creation of my wild beers and other fermented concoctions. We're talking wild berries, plants, and barks; fruit molasses, tree saps, or wild honeydew as sugar sources; leaves, roots, wild yeast, insects, and much more.

One of the fascinating aspects of what I do is the fact that very often I realize that I'm not really creating anything new; I've simply rediscovered some lost knowledge. A good example is the use of willow bark as a bittering ingredient for beer. I thought I had an original idea when I used it in some of my beers, only to find out quickly that it was used in traditional medicinal beers and even still is used in Germany to make a specific style of beer (Grätzer beer).

On the other hand, you can discover new possibilities from time to time. For example, in Southern California there are few if any recorded native uses of local plants for alcoholic consumption, but we do have some very interesting species such as California sagebrush (*Artemisia californica*) and yerba santa (*Eriodictyon californicum*). These are fantastic to flavor beers and were used medicinally as tea in the past.

California sagebrush is related to wormwood and mugwort, both herbs that were used instead of hops in older unhopped beer recipes. Yerba santa is a local aromatic herb that was used as a medicinal (cold and flu) tea. The flavor is really in the sticky sap, and if you boil it, you end up with quite a bitter tea. Because of these characteristics, both plants are used as bittering agents, like hops, in some of my primitive brews. You'll even find sagebrush and yerba santa beer recipes later on.

With a bit of research and experimentation, you can also create unique regional alcoholic drinks. By using what you have available in your vicinity, you're actually going back in time and rediscovering how the more refined and civilized beverages of today were invented in the first place. For example, the current near-obsession with using just hops and grains to make beer is somewhat of a modern anomaly in the history of brewing, which spans thousands of years—if not tens of thousands.

The quest to rediscover those long-lost flavors is an amazing journey. In his book *Uncorking the Past*, author and fermented beverages archaeologist Patrick E. McGovern made some fascinating discoveries about early alcoholic brews by analyzing residues from pottery. It seems our ancestors, in their quest for the perfect intoxicating libation, didn't hesitate to mix various brewing techniques, local ingredients, and sugar sources.

One of the earliest beers, over 9,000 years old and analyzed from dried residues in pottery at a Neolithic burial site in China, was found to have been made of mixed fermented ingredients such as grapes or hawthorn fruit, honey, and rice. I was absolutely thrilled to find a whole chapter devoted to early European brews, too. Analysis of pottery vessels and cauldrons from Scotland, Denmark, Spain, and other European countries revealed a similar approach to making inebriating drinks. Their mixing of grains, berries, honey, grapes, and bittering or flavorful herbs and spices made me feel much more confident regarding my unusual concoctions.

I found it very interesting that my use of mostly foraged ingredients to create my own brews resulted in the creation of, apparently, very similar drinks. In fact, many of the same plants I've used, such as mugwort, willow, yarrow, wild berries, and birch resin, were found in some of the residues that McGovern analyzed.

From my perspective the ancients' approach to brewing makes a lot of sense. You create with what you have and try to make it as flavorful as possible. It's really the basic rule for primitive brewing.

Just as I do now, ancient peoples created all kinds of brews with what could be found in their immediate environment; the concept of mixing all kinds of local ingredients to create a complex drink came naturally to them. If you think about it, raw honey is an awesome source of wild yeast, so it would be an obvious choice to mix into your brewing solution, especially if you lived in a colder climate such as Scotland, Sweden, or Denmark. It's also a good sugar source, albeit more limited and valuable than grains.

Maybe my own ancestors reached the same conclusion I did: There are no real rules! If it's enjoyable, somewhat tasty, and does the job, you've done your work!

On the Use of Wild Yeast

I often use my local wild yeast when making sodas, beers, wines, and other fermented beverages. I have done my best in my recipes to accurately convey the amounts of sugar and yeast used, which usually results in the specific alcohol content desired in the final beverage. However, because of the (wild) nature of what I do, you probably will end up with variations from time to time.

I don't see these variations as a burden; they're simply a representation and an intricate part of nature. Over time and through your own experimentation, you will establish a close connection with the peculiarities of your local terroir.

For example, the rule stating that wild yeast will ferment up to a certain point and then die (usually around 5 percent alcohol) is *mostly* true, but guess what? To this day, I continue to find exceptions. The yeast present on my local elderberries will produce a wine that can come close to 10 percent alcohol.

As a wild yeast source, raw honey is highly dependent on what can be found in the immediate environment. Over the years, I've had a couple of instances of beers or wines approaching 13 to 15 percent alcohol, which I attributed to the proximity of vineyards and the possibility of feral yeast—basically a takeover of a vineyard by the commercial yeast used to make wine.

From experience, I know that the same thing (a higher percentage of alcohol due to feral yeast) can happen if you make yeast starter from grapes purchased at the farmer's market or your local store.

So don't get discouraged if something doesn't work exactly the way you want it to the first time out. It's all part of establishing that special relationship with your local wild yeasts and their intricacies. If your yeast doesn't behave like you expected, work with it and feed it if necessary. Very soon you'll be able to tweak your recipes based on what you learn and create fermented masterpieces.

CHAPTER 2

Homebrewing Essentials

omemade brewing of sodas, beers, and wines is extremely easy, and you probably already have most of the equipment available in your kitchen.

If you're a novice and seek information online, it will seem at first to be a confusing and daunting task. For example, if you're interested in making beer or wine, you'll find all kinds of basic brewing kits (often composed of over 15 items), and you'll need to familiarize yourself with all kinds of exotic equipment and terms: hydrometer, bottle brush, specific gravity, mash pH, oxygen wash, iodine sanitizer, priming, cappers, and so on.

It's all good, but the reality is that brewing has been done for thousands of years, since well before most of those pieces of equipment, chemical sanitizers, or measuring tools existed. To this day brewing is still done using extremely simple tools in the desert of New Mexico (saguaro cactus wine), in remote villages of Africa (sorghum beer), and, heck, in my own kitchen.

Let's forget about all that complexity for now and go back to the basics. The biggest surprise that people have when they attend my workshops on primitive beer- or soda-making is how easy it is. The second surprise is how delicious some of the beers and other fermented beverages are. So let's look at what we need for basic brewing:

BASIC BREWING OR SODA-MAKING EQUIPMENT (FOR 1 GALLON/3.78 L)

Large pot (1½–2 gallons/5–7 L), with lid

Sieve

Couple of funnels

Measuring spoons

Scale

Depending on what you're brewing you'll also need a 1-gallon (3.78 L) glass bottle or jar (I recycled the bottle of a cheap Italian wine purchased at the grocery store)

The only things I tell people to purchase are:

Swing-top bottles (although you could use recycled soda bottles or
plastic beer bottles)
A small device called an airlock

That's it! Many of the projects in this book will require even fewer utensils.
Once you have the basic equipment, all you *really* need for this type of
brewing (beer, sodas, meads, and the like) is:

Water + Sugar Source + Yeast
+ Savory Ingredients (herbs, roots, bark, and so on)

Realize that each of the components necessary to brew already exists in your environment—or can be purchased. Your job is just to combine them properly so nature can do its magical work. Let's review some of these components.

Water

For my brewing projects I don't use my local tap water due to its chlorine content and, frankly, its flavor. I usually purchase springwater or purified water from the local water stores or the supermarket. I've also collected pure water from the local mountains and even melted snow during the winter.

Basically, the main rule I follow is to avoid water that contains chemicals that can inhibit my fermentation, as well as water that may contain chemical/organic pollution. As I've discovered from time to time, that beautiful stream in the forest can still have a dead raccoon in it, or other types of decomposing organic materials.

A few years ago, while living outside Los Angeles, I used well water with success. I've never used distilled water, but I imagine it would work, too. Some people argue that the lack of minerals in distilled water isn't good for the yeast, but if you brew with other natural materials, most will contain an adequate amount of mineral nutrients.

If you research online brewing forums looking for information about which kind of water to use, you'll encounter a lot of opinions. Factually, a lot of people are brewing with regular tap water.

It's your choice. If you're using a natural water source, just watch out for potential pollution (chemical and organic). Be aware that your water also affects the flavor of your fermented drink. I remember tasting some naturally sourced water in Croatia; it was so loaded with minerals that I could only describe the flavor as having "rock-like qualities." An acquired taste for sure.

In conclusion, I'll keep it simple: If the water is clean (free of chemical/organic pollution) and tastes good . . . use it!

Sugar Sources

Sugar can be found in all kinds of natural products, including sugarcane, sugar beets, fruits and berries, grains (malted grains), tree sap (maple, birch, palm, box elder, sycamore, and others), fruit juices, and even insect excretions (honey, lerp sugar).

When I became interested in primitive brewing using what nature was offering me locally, I thought that sugar would be hard to obtain. After a few years of research and experimentation, though, I've learned that this really isn't the case.

Here in Southern California you can make molasses or syrup by boiling down many local wild ingredients, such as cactus pears, figs, dates, various sweet berries, wild grapes, native plums, and cherries. If you're into urban foraging or gardening, the list of potential sugar sources really expands. Pretty much all year long, a quick visit to my local farmer's market provides me with sweet offerings—apples, pears, oranges, plums, cultivated grapes, apricots, and much more.

If you live in other regions or countries, your options may vary quite a lot, but it's been my experience that wild local sugar sources always exist. When I was in Vermont, maple or birch syrup was a readily available choice and provided a good representation of what the local wilderness has to offer. Molasses made with pears or sweet apples is another good option, and also makes an excellent addition to the maple syrup.

Of course, unless you live in the coldest or most arid regions of the world, honey is always available locally.

Originally, in my passion for exploring the flavors of my terroir, I tended to concentrate much more on foraged ingredients. These days I don't mind mixing stuff a bit more, and I often use common sugar sources such as brown sugar for brewing. That said, I recently made a fantastic "wild" molasses using local prickly pear cactus fruits and roasted lerp sugar (insect excretions), which I think make a terrific sugar base for a primitive brew.

Tips on Various Types of Sugar and Their Uses

Beer purists may crucify me for saying this, but it's a useful tip: If you aren't brewing with malted grains, brown sugar and commercial dark molasses (made from sugarcane or sugar beets) tend to make your brew taste like a beer. The same is likely also true of sweet sorghum molasses (which is mostly made in the South), but in all honesty I haven't tried that one myself yet.

By using brown sugar or molasses and bitter herbs such as mugwort, horehound, hops, or yarrow, you end up with something that will taste pretty much like a beer in the end. A lot of the people who drink my wild beers say that they taste like a cross between beer and cider. Maple syrup works great, too.

From my experience, refined white cane sugar is much better for primitive or country wines. It's often used if an additional sugar source is necessary

Some of the many sugar sources you can use for brewing.

to make the wine stronger because the original berries aren't high enough in sugar. Elderberry wine is a good example.

I also use white sugar to make my sodas. White sugar is very neutral and doesn't much alter the flavors of the plants used in the fermented drink. I don't think white sugar is fantastic for making beer-like fermented drinks, although there are records of it being used, in addition to malted grains, in some Belgian beers to boost the alcohol levels. But then again, one of my students used my mugwort beer recipe with organic cane sugar and, although it was closer to a cider, I was pleasantly surprised by the flavors.

Honey is essential for making mead (by definition a honey-based liquor), but can also be used in natural soda-type drinks and in beer. It seems to work well in both sodas and meads, but for a fermented drink that will have beer-like flavors, you still want to use malted grains, brown sugar, or molasses as the main sugar base.

Fermented beverages that use unusual sugar sources—molasses or syrup made from foraged fruits, insect honeydew, sweet juices, and so on—often taste like neither beer nor wine, but have their own flavorful peculiarities. Think of it as the Wild, Wild West of brewing, where "What the hell do I call that fermented concoction?" is the million-dollar question!

An alcoholic drink that I made using insect honeydew as a sugar base was a good example. It doesn't fit the definition or flavor profile of beer or wine, so I simply called it "Primitive Brew."

As you'll see later on in this book, there is a huge potential for creativity in making your own flavored syrups for brewing. By adding some unripe pinecone syrup to a basic recipe such as the mugwort-lemon beer, for instance, you can make your brew taste like the local mountains. And flavored syrups that incorporate wild or regular ingredients such as mint, rosemary, blackberry, wild fennel, or fig leaves can make great sodas.

Adding some tasty fresh or dehydrated sugary berries can elevate your wild beer.

Summary of Potential Sugar Sources for Fermenting Drinks

The following list includes my own preferred sugar sources based on my brewing experience, along with some rough guidelines for use. Feel free to experiment. For example, I'm sure you can find a way to make some tasty sodas with maple syrup; I just haven't done it using that sugar source . . . yet.

Molasses. Good for beer-type drinks.
Commercial brown sugar. Good for beer-type drinks.

Unrefined brown sugar such as panela/piloncillo. Good for beer-type drinks.

White granulated sugar. Use in sodas as well as primitive or country wines.

Syrups and flavored syrups. Excellent for soda-type fermented drinks or for adding a touch of a specific flavor to your beer.

Honey. Use for sodas, meads, and herbal meads, and also as an added sugar in beer-like drinks.

Insect honeydew. Use in primitive brews, and as an additive sugar in some of my wild beers.

Fruit molasses/syrup (reduced sugary juice). Great for sodas and primitive wines.

Dehydrated sugary fruits (grapes, wild currants, dates, figs, et cetera). Good in sodas and primitive wines, as well as to add flavors to your wild beers.

Tree sap or maple/birch syrup. Great for beer/cider-type drinks; birch syrup is also good for soda.

Agave syrup. Use in sodas, or as an additive sugar for beers and primitive-type brews such as *pulque.*

Palm sugar or syrup (also a tree sap). Use in sodas, or as an additive sugar for light beers and primitive-type brews.

Corn or rice syrup (neutral and bland). Although I haven't tried either of these, both have been used to make sodas or as a sugar additive for beer fermentation.

And of course you have **malted grain**, which is the stuff regular beer is made of.

I'm sure there are many more sources of sugar available; you just need to research and explore your local environment. I know John Muir (the naturalist and author) thought that sugar pine resin was better than maple syrup, for instance, but I haven't tried it myself yet.

You can tap many trees besides maple and birch for sugar. The process may not be as efficient, but it would make the sap or syrup that much more special. Examples of other tree species that can be tapped include box elder (*Acer negundo*), gorosoe (*A. mono*), butternut, aka white walnut (*Juglans cinerea*), black walnut (*J. nigra*), English walnut (*J. regia*), sycamore (*Platanus occidentalis*), ironwood (*Ostrya virginiana*), and possibly more.

Before attempting to tap any trees, do some thorough research and/or some apprenticeship (if possible) so you do it properly and don't damage the trees. Tapping trees is a beautiful craft in itself.

BROWN SUGAR—DID YOU KNOW?

Brown sugar either is natural unrefined sugar or is produced by adding molasses to refined white sugar. In the United States what we buy at the store is pretty much the latter. I can purchase true unrefined sugar at the local Hispanic market under the name *piloncillo* (in South America it's called *panela*). Of course, natural unrefined sugar has more sophisticated flavors than the common commercial product, but it might not be available in your area.

It's easy to make the same kind of brown sugar that you buy at the store. You may not save any money by doing it yourself, but it does give you more choices such as making very light or dark brown sugar. There are also some interesting applications, as you'll see.

How to Make Brown Sugar

1 cup (200 g) granulated cane sugar
1 tablespoon (15 ml) unsulfured molasses for
    light brown sugar, or 2–3 tablespoons (30–45 ml)
    unsulfured molasses for dark brown sugar

Mix together the sugar and molasses in a bowl. You could use an electric mixer, but I usually do it with a fork; it takes about 5 minutes. In the beginning, you'll have a bunch of globs, but as you keep mixing, eventually you'll end up with a beautiful brown sugar. Store in a ziplock bag or closed canning jar.

"Wild" Brown Sugar

Now that you know how to make brown sugar, you can also be more adventurous and create interesting sugars using local wild berries or what's growing in your garden. So far I've made "wild" brown sugar with blueberries, dates, and blackberries, and one of my favorites is elderberry sugar. As with making regular brown sugar, the technique is simple.

The first step is to make molasses with your berries (see page 29). You then use your own molasses in the same ratio as for making brown sugar:

1 cup (200 g) granulated cane sugar
1 tablespoon (15 ml) fruit/berry molasses for
    light brown sugar, or 2–3 tablespoons (30–45 ml)
    fruit/berry molasses for dark brown sugar

Follow the procedure described above for making brown sugar.

What about insects as a sugar source? We already use a common insect product for fermentation (honey), and I use lerps sugar, too, but what about the aphids that excrete honeydew or honeypots ants?

A bar in Brazil is presently making beer with lemongrass-flavored ants, which have been used as a traditional food source for eons. A few years ago I used local ants that tasted like lemons in my primitive brews, with excellent results.

And there are more unused or forgotten sugar sources you could research and explore in your local environment. Tree barks? Sweet edible flowers such as the ones from the eastern redbud tree (*Cersis canadensis*)? Roots or tubers like licorice, carrots, or sweet potatoes?

The possibilities that nature offers us are truly mind boggling. Our challenge is to explore and use these resources in an ethical and sustainable way.

Sugar Alternatives

Sweeteners that are non-sugar-based cannot really be used for fermentation. Stevia is a good example. Stevia is a natural sugar substitute extracted from a plant, but it doesn't really contain sugar. Because fermentation is a process of converting sugar into alcohol, a product like stevia would not work. You need actual sugar.

That said, adding stevia in the brewing process could be beneficial if you want an alcoholic drink or soda that is fully fermented (all the sugar has been converted into alcohol) but still has some residual sweetness in it.

Of course, there are all kinds of artificial sweeteners: aspartame (Nutra-Sweet), acesulfame potassium, saccharin (Sweet'N Low), Splenda, and more. I tend to stick to natural products, so I haven't used these products to sweeten my brewing concoctions. Because they don't contain any actual sugar, I also don't think they are suitable for fermentation.

How Much Alcohol Can I Get in My Fermented Beverage?

To be honest, achieving a specific alcohol content in my fermented drinks has never been a priority for me. I've always been interested in flavors foremost, but it's usually one of the first questions people ask when they drink one of my wild beers or wines: "How much alcohol is in it?" And my usual answer is: "I don't know, probably around such-and-such percent."

For me it's not that important to know exactly how much alcohol is in one of my brews. In addition, many of my drinks are live fermentations, so the alcohol content increases over time. A young beer may still be sugary and perfect for cooking, while an older one will be a bit more strong and sour.

I use a simple rule for my primitive drinks, which I call the "Neanderthal rule": 1 pound (454 g) of sugar in 1 gallon (3.78 L) of water will give you around 5 percent alcohol by volume. It's a very loose rule, but workable for many of the fermented drinks I make. We'll go over that rule very soon when we talk about yeast.

But to answer the question quickly (for now): A brew's alcohol content depends on the yeast and the amount of sugar. The more sugar is in your brew, the more alcoholic it can potentially become. In addition, some yeasts can create very alcoholic drinks (up to 20 percent), while other strains (including a lot of wild yeasts) cannot survive in high concentrations of alcohol and will stop the fermentation process when your brew reaches 5 or 6 percent alcohol.

You'll need to play with these two factors—sugar amount and yeast type—if you're interested in experimenting with alcohol content.

Flavored Syrups

Flavored syrups are great to experiment with. They make delicious simple sodas, but I like to use my syrups to add specific flavors to more complex drinks featuring seasonal plants and berries. This works especially well if something isn't in season; making all kinds of flavored syrups during the year allows you to be much more creative. A good example is elderflower syrup, which you can use to add floral qualities to your sodas, wines, or beers.

Making a syrup isn't complicated at all. There are no set rules, though usually for beer-like drinks I use brown sugar or molasses as a base, and for sodas I use white sugar or honey.

There are five types of simple syrup: very light, light, medium, heavy, and very heavy. The names are based on the amount of sugar used, a distinction that can be useful for food preservation (for example, fruits canned in syrup).

Types of Syrups

Type	Sugar %	Amount of Sugar Used
Very light	10	½ cup sugar per quart 100 g sugar for 1 L
Light	20	1 cup sugar per quart 200 g sugar for 1 L
Medium	30	1¾ cups sugar per quart 350 g sugar for 1 L
Heavy	40	2¾ cups sugar per quart 550 g sugar for 1 L
Very heavy	50	4 cups sugar per quart 800 g sugar for 1 L

MAKING SIMPLE INFUSED
OR FLAVORED SYRUPS

The number of delicious concoctions you can make with syrups is mind boggling. You can add wild or garden herbs, spices, and fruits (fresh or dehydrated), then use the syrups as your sugar source for fermentation, include them in cocktails, or simply add them to water for a refreshing drink.

The list of ingredients you can use is long: fig leaves, figs, lemons, oranges, limes, cinnamon, wild fennel, ginger, basil, mints, rosemary, lavender, vanilla, various sages, chile peppers, cactus pears, wild berries, spruce, white fir, unripe pinyon pinecones, juniper berries, and much more. One of my friends even uses vegetables such as celery.

The basic technique is quite easy: Place all your ingredients (plants, berries, barks, spices, sugar, water, and so on) in a pot and bring to a boil. Stir to make sure the sugar is dissolved, lower the heat, and continue simmering for around 10 to 15 minutes. Remove from the heat, let cool, and strain into a clean or sterilized container. Refrigerate for up to 3 weeks.

Some people like to simmer their ingredients for 15 to 30 minutes, and this can be a more efficient method for roots and barks.

For tender and aromatic herbs such as mint or basil, it's better to remove the syrup from the heat and steep the fresh leaves for 15 minutes or so, then strain and refrigerate.

Basic Berry Syrup

1 cup (200 g) sugar
1 cup (250 ml) water
1 cup (250 ml) blueberries or other
    berries (raspberries, wild currants,
    blackberries, etc.)

Combine the sugar, water, and berries in a small pot and bring to a boil. Stir until the sugar dissolves, and continue simmering for around 10 to 15 minutes. Let cool, strain into a clean (preferably sterilized) container, and refrigerate for up to 3 weeks.

If you know how to can, you can also check the acidity level, add some lemons or citric acid if necessary, and use the water bath method to preserve the syrup in sealed jars that can then be stored at room temperature until you're ready to use it.

Basic Herb Syrup

1 cup (250 ml) fresh mint or basil leaves
1 cup (200 g) sugar
1 cup (250 ml) water

1. Clean the herb leaves carefully. Combine the sugar and water in a small pot. Bring to a boil, stirring until the sugar dissolves, then remove from the heat. Place the herb leaves in the syrup and steep for 15 minutes.
2. Strain and pour the syrup into a sterilized glass jar or bottle. Let cool and refrigerate for up to 3 weeks.

Basic (Dried) Aromatic Herb Syrup

1 cup (200 g) sugar
1 cup (250 ml) water
1–1½ tablespoons (2–3 g) dried
   lavender blossoms

1. Combine the sugar and water in a small pot. Bring to a boil, stirring until the sugar dissolves, then remove from the heat. Place the blossoms in the syrup and steep for 15 to 30 minutes.
2. Strain and pour the syrup into a sterilized glass jar or bottle. Let cool and refrigerate for up to 3 weeks.

Those are basic recipes to get you started. As you can imagine, some fresh or dried herbs (rosemary, white sage, black sage, and more) can have very strong flavors. You will need to experiment a bit. My local wild sages are super strong, so syrup-making with them is really an exercise in moderation.

There are *tons* of flavored syrup recipes online using different amounts of sugar, longer infusing times, and slightly different procedures. Do a bit of research on your own and you'll be amazed by what you can create.

Once you get the hang of it, you can create some extremely interesting complex and delicious blends by combining herbs, berries, barks, roots, and so on. Much like making fermented beverages, syrup blends can represent whole environments.

FIG LEAF SYRUP

15 fig leaves
8 cups (2 L) water
Juice of 4 lemons
4 cups (800 g) white sugar or brown sugar,
    or 4 cups (1 L) honey

Procedure

1. Take your time and clean the fig leaves well. Meanwhile, bring the water to a boil, add the leaves and lemon juice, cover the pot, and simmer for 45 minutes to an hour.
2. Remove the leaves. Because of evaporation, you should end up with around 5 cups (1.25 L) of liquid. Choose the type of sugar you'll need for your future brewing: brown sugar for beer-like drinks, white sugar or honey for sodas, meads, and so on.
3. Add the sugar and simmer for around 15 minutes, then bottle.

HERB SYRUP

Feeling creative? Here in Southern California, I often add some local wild flavorful ingredients to my fig leaf syrup such as black sage, mugwort, yarrow, or pine.

2 cups (500 ml) water
2 cups (400 g) sugar
2–3 cups fresh mint or basil sprigs
    (12–15 fresh sprigs, or around
    2 ounces/56 g)
Juice of 1–2 lemons (optional)

Variations

Rosemary syrup: 6–8 fresh rosemary sprigs
    (around 1 ounce/28 g)
Thyme syrup: 6–8 fresh thyme sprigs
    (around 1 ounce/28 g)
Bay syrup: 14 bay leaves, or 4–5 California
    bay leaves (dried and aged 2 months)

Procedure

1. Pour the water in a pot, add the sugar, and bring the water to a boil, making sure the sugar is dissolved. Add the herb you've chosen and place it in the boiling water for a minute or so. Remove the pot from the heat. Add the juice of one lemon if you want (the acidity helps to extend the preservation time) and steep the herb for 20 to 30 minutes.
2. Pour the syrup through a strainer into a sterilized bottle or jar. Store in the fridge. With the lemon juice, it should keep for at least 3 to 4 weeks. Without the lemon juice, I try to use it within 2 weeks, but it could probably last longer. If you know how to can, you can always use the water bath method to make the syrup (with lemon juice) shelf-stable.

GREEN PINECONE SYRUP

Aside from regular syrups, as you research culinary or brewing uses for your local ingredients, you may end up with some unusual surprises. That's a big part of the fun.

Some syrups that can be used to add flavors in your brewing concoctions really don't follow the regular "Let's make some flavored syrup" methods. It's true for the Fig Leaf Syrup, but it's especially true for this recipe. I was doing some research about the uses of pine sap and how to collect pine nuts when I came across an obscure reference noting that the French were making a syrup with pinecones.

Until then, that interesting product had eluded me because the recipes were in French and pinecones were therefore called *pommes de pin* or "apples from pine." As soon as I Googled "sirop de pomme de pin," however, bingo! There were the recipes!

Pinecone syrup involves a fascinating procedure. The recipe was originally medicinal (used for cough and lung ailments)—it's part of the local herbalism tradition in France and Italy. Quite a few of the specific formulations don't even tell you which pines to use, simply calling for you to go out and pick up some green cones. Traditionally, and commercially, silver fir (*Abies alba*) and mountain pine (*Pinus mugo*) were used for their incredible flavors and possibly medicinal properties. I have also seen mention of black pine (*P. nigra*) and Italian stone pine (*P. pinea*). So far I have not found any tradition of making unripe pinecone syrup in North America, but you will find tons of recipes for pine needle teas, spruce and fir tip syrup, and pine needle syrup; these can help you determine which species of pine or spruce/fir trees to experiment with. I know that all true pines are considered "edible," but there are some reports of ponderosa pine having some abortive effects in large quantity (for goats). Lodgepole and Monterey pines are also sometimes listed as having medium toxicity, although pine needle tea has been made from pretty much all of them. So pick a local pine, check the flavor (chew on a few needles), and do a bit of research before using.

I'm completely in love with pinyon pine, and in the last two years I've been using the unripe green pinecones to make my syrup. The resulting flavor is nothing less than exceptional.

Unripe pinyon pinecone syrup in progress (I used brown sugar).

Note: Aside from making pinecone syrup, in Europe and Ukraine people also used to make pinecone jam, which involved using hot water and sugar. It's a bit outside the realm of this book, but by all means do some research on this if you're interested.

Procedure

1. Pick the green pinecones when they're loaded with sap. This is usually in late spring in California. Give them a quick, gentle cold wash if necessary. If collected in pristine condition, washing isn't a necessity.
2. Within a day or two after collecting them, place the cones in a jar and add brown or white sugar. Traditional recipes call for a ratio of 50 percent sugar to 50 percent green pinecones by volume. White sugar will give you a light amber syrup.
3. Close the lid and leave the jar in the sun for a month. The sugar will extract the water content/sap through osmosis and create the syrup. It's that simple! The California sun being so strong, I usually leave my jar in the sun for half a day and stop after 2 to 3 weeks based on the flavors.

Because I'd completed the Master Food Preserver program a few years earlier, I was initially a bit worried about the possibility of botulism given that I was leaving something in a closed container out in the sun for a month. Then I did a test on pinecone juice and found the acidity level quite high; this would inhibit botulism.

MAKING YOUR OWN
FRUIT SYRUP OR MOLASSES

Making molasses from sugarcane juice or sugar beets is a bit out of reach for most of us, but we can still have some fun and make some interesting versions using local ingredients, either wild or from the garden. As I explained earlier, I like to make fermented drinks out of nothing. By that I mean heading out into nature and making a fermented drink of what I find, from plants to sugar and yeast. The best representation of your own terroir likewise comes through using ingredients found only locally.

Sugar from sugarcane was imported to Europe after the Crusades. Before that, if you wanted to add sweetness to your dishes or drinks, you mostly used honey, which was somewhat expensive; tree sap in some regions (maple or birch syrup); or molasses made from sweet fruits and berries such as figs, carob, grapes, apples, pomegranates, and so on.

Many ancient alcoholic drinks are fermented with just the fruits and their sugar content. A good example is the wine made by the Tohono O'odham people in Arizona from the fruit of the saguaro cactus.

Even regular red or white wine is a good modern representation of an ancient drink; to this day some wines are fermented by simply using the grapes' sugar content and wild yeast from their skins in somewhat crude conditions. In the country of Georgia, following a technique and tradition more than 8,000 years old, Chinuri grapes, skins and stems included, are just left to ferment for months in sealed giant clay pots (*qvevri*) buried underground.

The technique to make fruit molasses is straightforward: Simply extract the juice from your fruits, then boil it to evaporate its water. You're left with a thick liquid mostly composed of sugar. Nowadays, because cane or beet sugar is so abundant, most modern recipes for making fruit molasses call for the addition of sugar to the juice, but not so in the original recipes. Some people, including me, add a bit of lemon juice or citric acid to make the fruit molasses more acidic and help it keep longer.

Is it a syrup or molasses? You probably can call it either, but it's often called molasses because, if prepared properly, it will have the same consistency as the regular molasses you can purchase at the store. If you don't boil it to the consistency of molasses, you can call it a syrup.

Lerp sugar: A lerp is a structure made of crystallized honeydew that protects the larvae of a psyllid bug. I collect the sugary lerps in July and August by scraping them off the leaves and dehydrating them in the sun. The honeydew is composed of roughly 60 percent sugar and 40 percent starch.

Modern Molasses

4 cups (1 L) fruit juice (from pomegranates, grapes, apples, etc.)
½–¾ cup (100–150 g) white sugar
Freshly squeezed juice from 1–2 lemons

Place the fruit juice, sugar, and lemon juice in a pot and bring the solution to a boil. Stir to make sure the sugar is fully dissolved, then reduce the heat to simmer the pot over medium heat. Supervise the process: In the beginning you may need to skim the top if you're using freshly made juice (as opposed to purchased fruit juice). Otherwise, the liquid may boil over the top. Depending on the type of fruits you're using, and their sugar content, the boil will take up to 90 minutes. You're basically looking for a reduced solution of around 1 cup (250 ml) with the same consistency as the regular molasses you purchase at the store.

You can do a plate test, similar to a jam test, by placing a bit of the hot liquid on a cold plate. (Put the plate in the freezer for about 15 minutes before trying this.) Place a spoonful of molasses on the plate and push your finger through it—unlike a jam, you're looking for it to flow back in, nearly reluctantly, to fill the gap. If it's not ready, keep simmering for 5 more minutes and test again.

There's no need to go crazy, though. If it looks and flows like molasses, you've done a great job.

Cactus pear molasses

Traditional Molasses

More traditional recipes use the same cooking technique to reduce the juice to molasses; the main difference is that you don't add sugar. It's basically just pure fruit juice, reduced and thickened.

The amount of molasses you can collect at the end varies greatly depending on how much sugar is in the juice in the first place. If you're a purist and plan to use the molasses fairly soon, you won't add lemon juice, but acid will definitely help preserve it longer.

Recently I foraged around 4 pounds (1.8 kg) of cactus pears, which gave me around 5½ cups (1.3 L) of juice. I followed the usual procedure for making molasses. Around 90 minutes later I produced a reddish black molasses that was more of a cross between molasses and jam in terms of texture. Because I didn't add sugar in the first place, I ended up with ⅓ cup (80 ml) of molasses. Delicious stuff, though!

By the way, there are no set rules or laws about making molasses. You could mix in fruits loaded with sugar such as pomegranates, figs, or dates. You can even add spices like cloves, cinnamon, and wild sages, or foraged aromatic plants including yarrow, mugwort, and wild mints.

A whole book could probably be written about mixing herbs, fruits, roots, barks, berries, and other ingredients to make fancy molasses.

Pascal's Fake Molasses Hack

If you don't have molasses at home and you need something with a similar consistency and flavor, you can make a substitute by mixing the following:

1¼ cups (300 ml) honey or maple syrup
1¼ cups (125 g) piloncillo sugar (commercial brown sugar works, too, but has less flavor)
¼ cup (60 ml) hot water

Apple Molasses— Did You Know?

It's nearly a lost tradition, but originally apple molasses was made by using fresh-pressed apple cider (what we in America call sweet cider—unfermented juice). The technique is simple: Simmer your cider until it begins to turn a dark amber color and reaches the consistency of a thick syrup (like honey). You're basically looking at an 80 to 90 percent reduction of the original liquid. Pour it in a sterilized jar or bottle and store in the fridge.

This makes me wonder if you could do the same with fresh-pressed perry (pear cider). Something to add to my list of experiments.

CHAPTER 3

In Search of Wild Yeast

Let's start with a simple definition: Yeast is a type of fungus that is used in making alcoholic drinks (such as beer, wine, or mead) and in baking to help make dough rise.

Basically, what yeast does is convert sugar into alcohol and carbon dioxide gas (CO_2). In fact, if you take a look at the origin of the word *yeast*, you'll discover that it comes from the Old English *gist* and Old High German *jesen* or *gesen*, which mean "to ferment."

Most people who are into making beers or sodas will purchase their yeast online or at their local brewing supply store. I'm not a commercial yeast expert, but basically the type of yeast that you use will imbue some specific flavors to your brew and also determine the percentage of alcohol you can expect after a full fermentation. I'm much more into what nature is offering me, so I tend to use wild yeast.

You may not realize it, but yeast spores are present everywhere. They're in the air we breathe; in plants, flowers, fruits, and soil; and even on our own skin. As I continue experimenting with brewing, I keep finding more and more sources of wild yeasts. For example, last year, while I followed an old recipe for unripe pinecone syrup, I found out (after my jar exploded) that my pinyon pinecones were completely loaded with wild yeast. So much so that I actually brewed some primitive beers by just placing unripe pinecones in the cold wort to ferment, and it worked like a charm. (*Wort* is a brewing term: It refers to the sweet infusion of herbs, ground malts, or other grains before fermentation, used to produce beer.)

When I taught a fermentation workshop at Sterling College in Vermont, I had the students experiment with wild yeast starters: They mixed some organic ingredients such as barks, leaves, flowers, and branches in sugar water. To my surprise, over 80 percent of the starters actually worked. Some of them were made with very unusual ingredients such as

Wild yeast starter using yellow birch bark and yellow birch branches with spruce tips.

Wild yeast starter using apple and dandelion flowers. Organic cane sugar.

Wild yeast starter using spruce tips and yellow birch leaves. (The latter didn't work.)

white pine branches, yellow birch bark, fern leaves, spruce tips, apple blossoms, and so on.

We ended up using a couple of starters that were super active (spruce tips and dandelion flowers). I decided not to use the most active one (fern) because, not being an expert about plants in Vermont, I didn't know enough about that type of fern. I found out later it was called sweet fern (*Comptonia peregrina*) and that it can be made into a tea. In retrospect, we could have used it. I also skipped the apple blossom starter when a student raised the question of cyanide (which is present in many fruit seeds, including apple, in small amounts). But further research indicated that it would not have been an issue. My philosophy is always to err on the side of safety. I also tell my students to start with something they *actually* know is not poisonous or unhealthy, and that a successful starter should smell good, too.

More than 1,500 types of wild yeast have been recorded so far, but thousands more probably exist in the wild waiting to be discovered and recorded.

Since the dawn of humanity, people have enjoyed a deep relationship with yeasts. In his book *Uncorking the Past*, author Patrick McGovern wrote about the first evidence of making an alcoholic beverage, which came from chemical analysis of jars from the Neolithic village of Jiahu in the Henan province of northern China. The results of the analysis confirmed that a fermented drink made of honey, rice, grapes, and hawthorn berries was being produced around 7000 BC.

Beermaking evidence has been found in present-day Iran on jug fragments that were at least 5,000 years old. And we also know the Egyptians made leavened breads using wild yeasts (sourdough) around 1000 BC based on hieroglyphs.

As science continues to improve and more ancient artifacts are retrieved, I'm sure we will find evidence of other peoples having made interesting alcoholic beverages much earlier in our history. Considering that the simple action of mixing raw honey (which contains a lot of wild yeasts) with water can create a natural fermentation within 3 or 4 days, it's highly likely that fermented drinks existed long ago in prehistoric times.

Until modern science came along, the invisible wonder of a liquid transforming itself into a magical solution capable of altering your senses was perceived as a divine intervention, often the domain of shamans and priests. In many cultures alcohol was considered a gift of the gods: Songs were written praising its virtues, and divinities such as Dionysus or Bacchus, god of the grape harvest and winemaking, were worshipped.

Primitive societies had many rituals revolving around this miraculous transformation. Some had rites that called for dances, chants, and noisemaking to attract spirits so they could inoculate the brew, while in other cultures calm and silence were the norms so as to not scare the spirits away.

The creation of wild brews can often be unpredictable. Such is the nature of wild yeasts, which are invisible and mysterious. At the same time the fermentations they create can be full of surprises and often delicious.

Primitive beers, wines, or even sodas made with wild yeasts can have ever-changing qualities due to countless factors such as the season, location, and yeast sources (plants, berries, and the like). While some people are nervous or disturbed by this lack of control, culinary adventurers see it as a blessing. Alcohol content, acidity, brightness, smoothness, and many other factors influencing the flavors will fluctuate in the same way nature does— sometimes unpredictably and esoterically. It is a true reflection of a wild, untamed terroir.

The modern trend in making beers or wines is to use specific commercial yeast strains that provide desirable and predictable results. For example, champagne yeast allows you to make fermented beverages with a high alcohol content. Some yeast strains can also be chosen for the flavors they impart, their speed of fermentation, or their ability to withstand low temperatures. If I ferment my wild beers using a commercial yeast, I tend to use a very plain dry yeast like Nottingham ale yeast. There is a *lot* to know about which types of yeast to use if you're an experienced brewer, but you can start with a basic yeast strain like the one I use.

If I don't use wild yeast (which is rare these days), I do the following:

Use beer yeast for my regular and wild beers.
Use champagne yeast for sodas.
Use wine yeast when fermenting wild berries, grapes, elderberries,
   and the like.

Don't use a bread yeast for brewing beers or sodas, as the end result will have bread-like flavors. I guess it could be a good thing if you really like bread, but I don't find the flavor very appealing in a beer-like drink.

Commercial yeast strains are also designed to achieve higher levels of alcohol. Results will vary somewhat, but as a general rule commercial beer yeast can give you around 10 to 12 percent alcohol, wine yeast around 12 to 13 percent, and champagne yeast 13 to 15 percent. And of course specific types of yeast can give you a much higher percentage of alcohol, such as White Labs Super High Gravity yeast (over 20 percent alcohol).

The old trend (which by the way is becoming popular again) is to experiment with what nature has to offer, but wild yeast usually doesn't have the same tolerance for higher levels of alcohol as commercial yeast. Most wild yeast will die at around 5 to 6 percent alcohol, although I've had some wild yeast going higher, such as the yeast from my local elderberries.

Presently, I would say that 75 percent of my brews are made with wild yeasts. Late spring and summer are the best times to forage these microorganisms, as they are attracted to sugars and we have a lot of berries ripening in California during those seasons—elderberries, wild grapes, and juniper berries, to name a few. These days I'm also completely in love with unripe pinyon pinecones as starters.

The Neanderthal Rule

A few pages ago I wrote about the Neanderthal rule for calculating the amount of alcohol you may get in your brew. If you don't use malted grains, the basic rule is that 1 pound (454 g) of sugar in 1 gallon (3.78 L) of water will give you around 5 percent alcohol, assuming that most of the sugar is fully converted into alcohol.

It's a very loose rule, but it works pretty well for white sugar, brown sugar, molasses, and even honey. It's a bit different and more difficult to figure out when you start using homemade flavored syrups for making sodas or other alcoholic drinks.

There's nothing wrong with trying to evaluate in a much more scientific way how much sugar you have to start with in the liquid when you make your beer or soda (this is called *initial gravity* in beermaking). You can purchase an instrument called a hydrometer at your local brewing supply store or online; it's easy to use.

I actually verified the Neanderthal rule using a hydrometer. Assuming all the sugar is converted into alcohol, you should be able to obtain the following final alcohol percentages:

> 1 pound (454 g) honey in 1 gallon (3.78 L) of water gives
> 4.8 percent alcohol by volume (usually abbreviated as ABV).
> 1 pound brown sugar per gallon gives 5.2 percent alcohol.
> 1 pound piloncillo sugar per gallon gives 5.3 percent alcohol.
> 1 pound molasses per gallon gives 5.3 percent alcohol.
> 1 pound white sugar per gallon gives 5.2 percent alcohol.
> 1 pound maple syrup per gallon gives 5 to 5.3 percent alcohol.
> 1 pound dry malt extract per gallon gives 5.2 percent alcohol.

Results may vary a bit depending on the brand of sugar or molasses or the type of honey used, and not all sugar will be converted to alcohol. But as you can see, the Neanderthal rule is pretty good.

If you use flavored syrup as a sugar source, use the following rule:

3 cups (700 ml) very heavy syrup (using white sugar) in 0.85 gallon
(3 L) of water gives around 5.5 percent alcohol by volume.

Thus, by deduction, for a syrup made with white sugar, you should end up
with the following:

3 cups (700 ml) very light syrup in 0.85 gallon (3 L) water gives
around 0.6 percent alcohol.
3 cups light syrup in 0.85 gallon gives around 1.2 percent alcohol.
3 cups medium syrup in 0.85 gallon gives around 2.1 percent alcohol.
3 cups heavy syrup in 0.85 gallon gives around 3.3 percent alcohol.

If you use a syrup made with honey, you'll end up with a bit less alcohol.

The reason it's good to know the Neanderthal rule is because as you
experiment with wild yeasts, you'll realize that you may have to tweak your
regular recipes. For example, my usual mugwort beer recipes use 1¼ to
1½ pounds (567–680 g) of brown sugar per gallon. If I use wild yeasts,
though, I usually only use 1¼ pound (567 g) of sugar, because wild yeast
does not survive well in high alcohol concentrations and usually dies off
once the beer reaches around 5 percent alcohol. Using the regular recipes,

Wild yeast starter using organic grapes. Day 3.

1½ pound (680 g), my wild yeast brew tended to be too sugary because the yeasts died and a lot of sugar was not converted into alcohol.

The key to foraging wild yeast is to find it in enough quantity to make a good starter. I tried a few times to make wild beers by placing some of my cooled herbal solution outdoors, in the hope that yeast spores present in the air would somehow decide to make a home in it. But I had no success with that method. After a few days, I would see mold form on the surface of the solution, and the smell became unpleasant enough that I didn't want to risk drinking any of it.

Like anywhere in this world, it's a competition out there. You have all kinds of microorganisms all competing for survival. Even if a few yeast spores land in your brew, there may not be enough of them to win the battle against other organisms that would spoil it. The key is to find yeasts in quantities that actually give them an edge and help them take over.

If you decide to forage yeasts, you'll have the best chances with flowers, berries, and fruits. All are usually loaded with sugar, and very often you will notice a white "bloom" around them. It's very similar to the bloom you see on organic grapes if you go to the farmer's market. This bloom is actually wax, but it contains a lot of yeasts. My local elderberries and juniper berries have a substantial amount of bloom, which helps a lot in making some good starter for my wild brews.

Good Sources of Wild Yeasts

Grapes, plums, fruits that have a white bloom. From your garden or farmer's market. Make sure they're organic.

Gingerroot. This must be organic—most of the ginger I find locally is imported from China and doesn't work for fermentation. I suspect it's been irradiated to kill potential microorganisms and insects.

Fresh wild juniper berries. Note that not all juniper berries are edible.

Elderberries

Wild grapes

Elderflowers

Blueberries

Blackberries

Figs

Prickly pear fruits

Tree barks—birch (*Betula* spp.) and aspen (*Populus* spp.). Yellow birch bark (*Betula alleghaniensis*) works, too, but I think it should be collected in late spring or early summer.

Unripe pinecones. My pinyon pinecones were loaded with yeast, and many people have reported excellent results using unripe pinecones from their local pines. You'll need to experiment a bit. Pinyon pine or white pine branches. It's likely that other species of pines are suitable as well, but I haven't experimented with them yet.

Raw honey

A lot of unwashed organic fruits (apples, peaches, lemons, et cetera) are also excellent sources of yeast. Make sure they're organic and clean (not grown in a polluted environment and sprayed with chemicals).

How Much Starter Do You Need for Brewing?

I usually use around ½ to ¾ cup (120–180 ml) of yeast starter per gallon (3.78 L) of beer or soda. Some recipes I've seen online use 1 cup (250 ml) per gallon. The consensus seems to be somewhere between ½ and 1 cup.

Note that some yeast starters, such as pinyon pine or juniper berries, may add flavors to your brew. This isn't excessive, though. You can create some yeast starter mixes specifically to accentuate or add character.

Unlike commercial yeast, wild yeast may still require 2 to 4 days to take off in your wild beer or soda. If the recipe calls for 10 days of fermentation, you would start counting the days when the fermentation is quite active.

I've also been successful with getting some of my wild brews to ferment by simply adding ingredients loaded with wild yeast into the liquid (wort), such as pinyon pinecones or juniper berries. I remove them when the fermentation is doing well.

Wild Yeast Summary

If you're new to making beers, sodas, or wines, I advise you to start with cultured commercial strains of yeast. It's much easier for you to learn; try various recipes and achieve success the first time around. Also, if you're into making alcoholic beverages with high levels of alcohol, this is probably your best bet. Regular brewing yeast (beer, wine, or champagne) can be purchased in liquid or dry form at a local brewing supply store or online.

As you gain experience, or if you are already an experienced brewer, start to play with wild yeast—a fascinating and truly enjoyable experience. There is something special about being able to create an interesting alcoholic drink using only hyperlocal ingredients from your own garden or the local wilderness. I believe I can truly taste the local terroir in some of my brews.

Besides, going foraging for and experimenting with wild yeast sources and making starters is such a fun activity. If you're a teacher, it's a great way to educate kids on nature's wonders.

Pick wild berries or fruits that have a nice bloom. The waxy bloom contains a lot of yeast.

Place around 15 to 20 percent sugar (or honey) and 85 to 80 percent water (not tap water) in a jar.

Add the berries to the sugary solution.

Close the lid, but not too tight. Three or four times a day, close the lid tightly and shake, then unscrew the lid a bit again.

Active fermentation from California juniper berry starter. Day 4.

The Wildcrafting Brewer

MAKING A WILD YEAST STARTER

The idea with making a yeast starter is to create a solution where you increase dramatically the cell count of the yeast by feeding it with sugar before "pitching it" (pouring it) into your brew. The large quantity of yeast pretty much ensures a successful fermentation.

The method I presently use is almost 100 percent effective.

Procedure

1. Make a sweet solution composed of 15 to 20 percent sugar and 80 percent water by volume. Don't use tap water, which may contain chlorine. Honey can be used, too. If the liquid tastes quite sugary, you should be fine.
2. Place the solution into a clean bottle or jar. Pasteurizing the container by placing it in boiling water for more than 10 minutes is even better and will increase your chances of success, but in my experience it's not a must. Washing with soap and water also works well.
3. Place your berries (or other foraged yeast sources) in the container with the sugar solution. I don't have a precise quantity, but it's usually around 20 to 30 percent of the solution or even more.
4. You don't want flies or unwanted bacteria to contaminate your starter, but you need to let fermentation gases escape, so tie a clean kitchen towel or paper towel—or even better, place an airlock—on top if you're using a bottle. For jars, you can use a regular lid and band—just don't screw it on too tightly. You want to let fermentation gases escape.
5. Shake the container or use a clean spoon to stir the contents three or four times a day. If

you're using a jar, screw the lid tight before shaking, then unscrew it a bit again afterward. Around 2 to 4 days later (less in hot weather), you will notice some bubbling in the solution. Congratulations: Your fermentation is active. Just in case, I like to smell it, too, as added insurance. If it smells really bad, don't use it.

6. Add some of the fermenting starter to the liquid you want to ferment. I use around ½ to ¾ cup (120–180 ml) of starter for 1 gallon (3.78 L) of brew. If you have a recipe asking for 10 days of fermentation, start counting the days once you have achieved an active (nicely bubbling) fermentation. It may take 1 to 3 days to get a fermentation going with a starter.

Note: As you'll learn later, mostly when making sodas, herbal meads, or even primitive beers, you don't always have to use a starter. Sometimes I just let the wild yeast in the ingredients do its thing. A yeast starter is mostly useful when you've had to boil your ingredients to extract flavors (and thus killed off any already existing yeast).

Wild yeast starter using unripe pinyon pinecones. Notice the bubbles; I had just unscrewed the lid a bit.

Wild yeast starter using pinyon pine branches and needles.

Wild yeast starter using cactus pears.

Wild yeast starter using elderflowers.

Wild yeast starter using elderberries.

Brewing without Using a Yeast Starter

Having an active yeast starter isn't always a must, but it's nice to use as an insurance that your fermentation will start well.

In Belgium some traditional breweries use a process called *open fermentation* whereby the boiled and strained sweet solution of plants and grains (the *wort*) is exposed to the air in what's known as a *coolship*. This is basically a large, shallow vat used to cool the wort after the boiling phase. As the wort is cooling, ambient yeast and bacteria found in the air "infect" the brew and start the fermentation process.

Some of these breweries have used this process for so long that the yeast inhabits the very structure of the building, from walls to roof. One Belgium brewery experienced trouble with their open fermentation process during a renovation of their facility. Due to modern civilization and the destruction of the local wilderness through urban expansion, the wild yeast that had originally been present in the air a century ago was going through an extinction process. Unbeknownst to the brewery staff, they were actually relying on the building itself and its invisible yeasty ecosystem to continue the traditional brewing process.

Here in Southern California I haven't experienced much success with the open fermentation method. I think it's probably because the temperature here is usually quite hot, even during the winter, and having a large, shallow vat outside without protection isn't really an option given the numbers of fruit flies and other critters flying about during most of the year.

My usual experience has been that a wort goes bad after a few days, and based on the smell, it's definitely not something you would want to drink.

An open fermentation may be an option if you live in a cooler climate, or it could be a possibility during the colder months in Northern California, but to be honest I haven't tried very hard to use this method given the abundance of wild yeast on local fruits, flowers, and berries.

My usual way to bypass the process of creating a starter is simply to place into my cooled wort some ingredients that I know are loaded with wild yeast. Good examples include our local juniper berries, pinyon pinecones, elderberries, and so on . . . but you could also use the skins (preferably featuring a beautiful bloom) of organic fruits such as plums or grapes. After 2 to 4 days, the fermentation is usually active and I remove my yeast-providing ingredients unless I'm interested in the flavors they can contribute to my brew.

For this type of fermentation, I usually use a large clay or glass container with a clean towel on top, or a plastic bucket equipped with a lid and airlock. The buckets can be purchased online or at your local brewing supply store. I use 2-gallon (7.6 L) buckets, but you can find some that hold up to 20 gallons (76 L).

Brewing without a yeast starter isn't my favorite approach to fermentation. For one thing, it's a bit riskier, but I've done it quite a few times nonetheless just for the fun of it or because I didn't have any commercial yeast or wild yeast starter at home. It is a valid method and worth experimenting with. There's something quite primal and exciting about the process of throwing flavorful ingredients together and just letting nature do its thing.

Unripe pinyon pinecones were placed in a cooled wort.
Fermentation started after a couple of days.

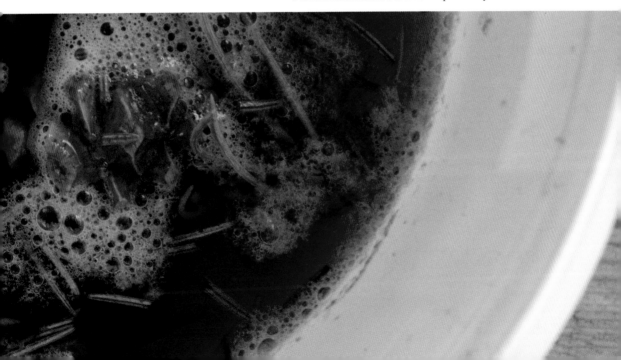

GINGER BUG

Another way to get wild yeast while skipping the foraging part is the famous ginger bug. As the name implies, it's basically a starter composed of ginger-root and the yeast culture inhabiting it (the "bugs"). It's one of the best-known ways to get a starter going, and if you do some research in books or online, you'll find tons of recipe variations and opinions on how best to go about it. What's important is that it works. I'll share my own method here, but by all means do a bit of online searching and pick a method that you like, or experiment and develop your own.

As you probably know by now, I like to make yeast starters with wild berries, fruits, or flowers, but there's nothing wrong with going to the farmer's market and purchasing some *organic* gingerroot. A ginger bug is actually supposed to be a mix of wild yeast and lactic acid bacteria (I think it's mostly wild yeast); like vinegar or kombucha, it can be continuously fed and kept alive. I'm a bit late to the ginger bug game, but so far I've been using my jar for a few months and I've experienced no issues.

It's super easy to start. Note that non-organic ginger (often imported from China and sold in the United States) may not work due to possible irradiation, which kills all bacteria.

Here is how I go about making my own ginger bug.

3–4 tablespoons (40–50 g) grated/sliced organic gingerroot, skin on

½–¾ cup (100–150 g) white sugar or brown sugar (I prefer white sugar for sodas); honey is okay, too

2 cups (500 ml) water—don't use tap water, which may contain chlorine

Procedure

1. Mix all your ingredients in a clean canning jar (I use a quart jar). Screw the lid on top, but not too tightly; you want fermentation gases to escape. Three times a day, screw the lid on tight and shake vigorously for a few seconds, then loosen the lid again.

2. Twice a week, add 1 tablespoon (15 g) grated or finely chopped gingerroot, 1 tablespoon (12 g) sugar, and 1 tablespoon (15 ml) water. Some recipes ask for 2 tablespoons (30 g) ginger, 1 tablespoon (12 g) sugar, and 2 tablespoons (30 ml) water. As you can see, the recipe is quite flexible.

   Usually it takes around a week to get a nice bubbly fermentation going, but here in Southern California I can get it going in 2 to 3

days during summertime. If the fermentation doesn't happen, you'll need to start again. It's probably not your fault, and the ginger may not have been organic.

If you're successful, you can keep the bug alive by continuously feeding it: Add 1 tablespoon (15 g) of grated/chopped root, water (15 ml), and sugar (12 g) every 3 to 4 days or so. If you don't make fermented drinks every few days, though, that biweekly chore is not a necessity and you can decide to let it "sleep" in the fridge and feed it once a week with the usual amount. When you're ready to use it again, bring it to room temperature to reactivate the yeast and start the twice weekly feeding ritual again the following day.

To be honest, I'm very, very lax with this method. I've sometimes left my ginger bug in the fridge without feeding it for 3 weeks and experienced no issues. So you have a decent amount of leeway in managing your bug.

To use ginger bug as a fermentation starter, simply strain and add around ⅓ cup (80 ml) of starter liquid for every ½ gallon (1.89 L) of juice, sweetened tea, or other sugary concoction you want to ferment. Pour the final liquid into a recycled soda bottle or Grolsch-style swing-top bottle, let it ferment for a couple of days, then place it in the fridge. In Los Angeles my drink can be ready in a day thanks to the temperature, so make sure to check how your fermentation is going.

After you're done using the ginger bug, place it back in the fridge until you're ready to start a new fermented creation. If you see any mold growing in it, discard the ginger bug and restart the whole process. When you have excess ginger bug, just use it for composting.

CONTINUOUS WILD (NON-GINGER) BUG

The problem I have with ginger bug is that I can't find ginger during my foraging trips, and I have this strange obsession about using my own local wild yeast to ferment my sodas, beers, and wines. So I decided to create my own continuous "wild bug."

It's similar to making a ginger bug: You get a fermentation started with one or more yeasty ingredients and keep adding sugar, water, and various yeast sources as they become available during the year to keep the fermentation going. The idea is to keep your wild bug alive and healthy by introducing new yeast strains, food (sugar), and water. So far, I've had my wild bug going for a bit more than a year.

My favorite local sources of wild yeast are:

California juniper berries
Prickly pears
Blueberries
Unripe (green) pinecones
Elderberries
Elderflowers
Pinyon pine branches
Wildflowers

The yeasty ingredients I place in the sugary water change with the seasons. At some point I need to remove some of the old ones while I add the new ones.

Procedure

The method is pretty much identical to that for a ginger bug. Place the ingredient(s) coated with wild yeast in a sugar solution composed of around ½ cup (100 g) white sugar and 2 cups (500 ml) water. Don't use tap water, which may contain chlorine. I usually use a small handful of berries to start with, two or three prickly pears, or a couple of unripe pinecones. Experiment a bit with your local source of wild yeast.

If you used a quart (1 L) jar, screw the lid on top but not too tightly, so that fermentation gases can escape. Three times a day, screw the lid on tight and shake vigorously for a few seconds, then loosen the lid again.

In 3 to 4 days the fermentation should be going well. At that point start adding 1 or 2 tablespoons (12 g) sugar and 2 to 4 tablespoons (30 ml) water once or twice a week.

If I don't have an immediate use for it, I place the jar in the fridge and feed it once every week or two with the usual amount of sugar and water (while it's still cold). When I'm ready to use it again, I'll bring it to room temperature to reactivate the yeast and start the regular feeding ritual again the next day.

These days I'm very loose with the feeding ritual and go by smell/taste to determine if I should feed it or not. I sometimes use as much as ½ cup (120 ml) of sugary water if I feel it's necessary, or sometimes just add sugar if there's already a lot of liquid in the jar. If you do a lot of wild fermentation, the process becomes instinctive; you just know what to do based on prior experience. But it's good to follow some rules in the beginning.

My wild bug follows the seasons: The juniper berries will be replaced by unripe pinecones, which in turn may be replaced by elderberries and then prickly pears. If the jar becomes too full, I remove 60 percent (or more) of the liquid and old ingredients and add new water and sugar as well as new yeast sources.

The liquid will also change color during the year. In late summer and fall, it's usually a dark red color due to my use of elderberries and cactus pears as yeast sources.

Wild ingredients coated with wild yeast are added and removed during the year to make this continuous wild bug, which will be used as a wild yeast starter for my beers, sodas, and wines.

CHAPTER 4

The Quest for Flavors

A s I was planning this book a couple of years ago, I thought I'd include a section listing pretty much all the herbs, plants, barks, berries, and other natural ingredients that people have used in the past (and present, too) to make sodas and brew beers, meads, spiced wines, or other alcoholic beverages. I was particularly interested in plants used instead of hops or as additional bittering agents in ancient beers.

But something interesting happened along the way: The more research and experimentation I did, the more I came to the conclusion that trying to list everything is pretty much an insane and impossible task. Why? Because, again, there don't seem to be any rigid rules as to what you can or cannot use, and the more I studied, the more I realized that since the dawn of humanity (and in fact until very recently) people have used whatever they had in their environment to create their alcoholic drinks, from plants to flowers, barks, fruits, berries, leaves, and more.

Let's take a look just at plants. Granted, not all of them are edible or healthy to consume, but scientists estimate there are over 400,000 plants on earth—that's a lot of flavors to play with. Here in Southern California, I always tell my students that I can use around 80 percent of what's in my environment for culinary applications, so even if just 20 percent of all the plants on earth could be used in some way for cooking or brewing, that's more than 80,000 potential ingredients. Of course, some of them will have similar flavor profiles—say, grasses, with over 10,000 species—but overall there is still an incredible number of possibilities. Then we have berries (estimated at over 7,000 species) and fruits (estimated between 2,000 and 3,000 species). In addition, some fruits come in many varieties; for example, you'll find over 7,000 varieties of apples grown in the world.

Ingredients collected at my friend's place in the local mountains and used to brew a "wild" beer: mugwort, yarrow, unripe manzanita berries, willow bark, pinyon pine branches, and unripe pinyon pinecones.

If you add other natural, organic ingredients that are not considered plants but have been used in brewing, such as lichens, mushrooms, and algae, you've added around 20,000 more possible species. Again, most of them may not be edible or healthy for you to consume, but even if only 5 percent could be used, that's another 1,000 potential flavors! Locally I use turkey tail and candy cap mushrooms, but quite a few beers and fermented beverages have included mushrooms such as chaga, reishi, black trumpets, maitake, shiitake, and many others for their medicinal qualities, their earthy flavors, or as bittering agents.

Just to give you an idea, the website Untappd (untappd.com), which lists beers from all over the world, gives more than 100 beer results if you search for "mushroom," 965 beer results for "spruce," 13,538 beer results for "pumpkin," 55 beer results for "kelp" . . . the list goes on.

We're not even touching upon insects (close to 1,000,000 species worldwide)—which, yes, have sometimes been used in brewing as flavoring or coloring agents. In South America a beer is made using lemony-tasting ants, and bitter-tasting cochineal bugs have been used as a coloring agent in Campari and various beers. I have experimented locally with using my own lemony-tasting ants, with great success. Insect excretions such as honey or honeydew have a place in the brewing process as well. I use my local lerps sugar (insect honeydew) in quite a few primitive beer recipes. Even the beer you purchase at the store contains some bug parts, too, as it would be impossible for a large brewing company to completely eliminate all bugs such as beetles and weevils from the grains they use.

I could go on and on detailing all the potential ingredients, but my point is really that we're dealing with an almost infinite number of possibilities.

Even flavors haven't always been the main goal through history. These days we seem a bit obsessed with creating the perfect wine, or brewing an awesome-tasting beer or soda. But we have also taken the medicinal, sacred, and religious context out of the process. If you start experimenting with truly primitive or sacred brews, you'll soon realize that some of them were drunk not for their flavors but for the psychotropic effects of the alcohol or the ingredients themselves.

Presently we know and understand what yeast is and how fermentation occurs, but for many older cultures fermentation was seen not as a natural process but as a divine or spiritual intervention. Making a sacred drink could include elaborate ceremonies, noise, and songs to attract the yeast spirits; other cultures felt serenity and calm were required so as not to frighten them away. Fermentation was a spiritual act in and of itself.

From my perspective, it truly is a sacred process. Think about it: Billions of tiny life-forms are elevating a somewhat ordinary drink into an intoxicating libation that can free your spirit, alter your senses, ease pain, and make you happy for a while. Pretty amazing stuff!

If you're an herbalist or inclined to brew for health reasons, flavors would not be your main goal. Back in the Middle Ages, drinking water was an iffy proposition—especially in cities—due to organic pollution and the potential of poisoning from *E. coli*, salmonella, listeria, and other bacteria. Drinking low-alcohol beer, ale, or wine was a much healthier alternative thanks to the pasteurization that often occurred in the heating process prior to fermentation. A few months ago I re-created an ale from the Middle Ages based on a recipe in the book *Ale, Beer, and Brewsters in England* by Judith M. Bennett, and it was a fascinating experience! You could not compare that ale to what you can presently buy at the store; it reminded me more of a sort of highly nutritious liquid porridge with barely any noticeable alcohol in it. From the perspective of my modern palate, I had a hard time comparing it to modern ales.

Aside from nutritious liquid for your body, specifically chosen herbs or barks could be added for their medicinal qualities. For example, willow bark acts a lot like aspirin and can reduce pain and fever.

Not just beers or ales, but all kinds of fermented drinks were created for the purpose of addressing specific health issues. Nettle beer is a good remedy for scurvy, and dandelion wine was used as a digestive.

But getting back to the subject: Are there any rules when it comes to flavors?

Sort of. Bitter flavors are for beers, sweet and fruity for sodas. Wine flavors are more difficult to put into words, but you're looking for specific qualities such as "full-bodied, fruity, spicy, expressive of terroir." And of course there are exceptions for every single type of beverage. You'll find bitter or unusual sodas such as San Pellegrino Sanbitter, or mauby soda made with bitter bark. Belgium has many sugary and fruity beers, like Rodenbach and Framboise Lambic. And many homemade wines with local fruits or wild grapes don't even remotely compare to what you can purchase at the store.

Then you can go even further and try your hand at making fermented drinks that defy the easy labels of beer, soda, cider, and wine. Some examples are *smreka* (fermented juniper berries), *kvass* (a fermented beverage usually made from black or regular rye bread), or my own alcoholic beverage made with insect honeydew (lerps sugar). In fact, many of my drinks made with foraged ingredients fall into a murky classification zone.

After years of studying nature and experimenting with her gifts, while I fully recognize that very specific plants have been used extensively in the past, I also came to the conclusion that many people simply experimented with what their local terroir had to offer when they created their own unique libations. What it means for us today is that we have tremendous room and freedom for creativity.

Mugwort (*Artemisia vulgaris*)

The Bitter World of Beers

If you're reading this book and are more interested in making sodas, fruity beverages, or herbal meads, feel free to skip this section and go directly to the information about soda (page 69). However, I think knowing about some of the herbs that were used to make herbal beers can add to your flavor arsenal when you're preparing sodas or meads, too. For example, one of my favorite combinations in soda is mugwort, pinyon pine, and pears.

As a general rule, bitterness is what you'll find in beers. You don't have to be a brewing expert to make that judgment; just grab a few beers from your local supermarket or liquor store and taste. From my perspective, bitterness complements nicely the somewhat sweet taste of fermented grain.

The exclusive use of hops as a bittering, flavoring, and aromatic agent in beer or ale is somewhat of a modern phenomenon in the history of brewing. The first recorded use of hops in beer is from the 11th century, although there is a mention of hops cultivation in the 9th century. At the time, hops was just one of the many bitter ingredients used to flavor beers, including mugwort, yarrow, ground ivy, dandelion, rue, and many others.

Today alcoholic beverages made with ingredients other than hops are called *gruits*, but at the time they were simply considered beers.

It's way beyond the scope of this book, but I really encourage you to read about the history of brewing. It's quite entertaining if you have a good author—very much like watching a soap opera. Lots of intrigues, politics, and religion are involved, plus the creation of monopolies and of course . . . money! Deciding on which ingredient(s) must be used, as required by law, to create a beverage called beer meant that you could also tax it. Nothing new, really.

In the 15th century, the Germans came up with the *Reinheitsgebot*, which is basically a set of regulations that set the price for selling beer and permitted only water, hops, and malt as ingredients.

The actual text reads as follows:

> We hereby proclaim and decree, by Authority of our Province, that henceforth in the Duchy of Bavaria, in the country as well as in the cities and marketplaces, the following rules apply to the sale of beer:
>
> From Michaelmas to Georgi, the price for one Mass [around 1 liter] or one Kopf [a container holding around 1 liter of liquid] is not to exceed one Pfennig Munich value, and

From Georgi to Michaelmas, the Mass shall not be sold for more than two Pfennig of the same value, the Kopf not more than three Heller [worth around half a Pfenning].

If this not be adhered to, the punishment stated below shall be administered.

Should any person brew, or otherwise have, other beer than March beer, it is not to be sold any higher than one Pfennig per Mass.

Furthermore, we wish to emphasize that in future in all cities, market-towns and in the country, the only ingredients used for the brewing of beer must be Barley, Hops and Water. Whosoever knowingly disregards or transgresses upon this ordinance, shall be punished by the Court authorities' confiscating such barrels of beer, without fail.

Should, however, an innkeeper in the country, city or market-towns buy two or three pails of beer (containing 60 Mass) and sell it again to the common peasantry, he alone shall be permitted to charge one Heller more for the Mass or the Kopf, than mentioned above. Furthermore, should there arise a scarcity and subsequent price increase of the barley [also considering that the times of harvest differ, due to location], WE, the Bavarian Duchy, shall have the right to order curtailments for the good of all concerned.

Although resisted in some parts of Europe (like England), over time the notion that the beverage called beer could only be made with hops, grains, and water was accepted . . . and taxed.

Why hops were chosen as one of the main ingredients is highly debatable, and there probably isn't one right answer. A set of geographic, political, religious, and monetary circumstances all came into play, in addition to the fact that hops is a quite flavorful bitter herb and has antiseptic and preservative qualities that help prevent spoilage.

Interestingly enough, hops is actually a perennial plant of the Cannabaceae family, which also includes the genus *Cannabis* (hemp, marijuana). From an herbalist's perspective and among its other properties (antibacterial, digestive, anti-inflammatory, et cetera), hops is also a sedative, and mixed with alcohol it's a very effective painkiller.

If I was into controversies, I would say it's interesting that an herb that puts you to sleep and eases pain was chosen as the one people should have in their beverage, but I'll leave it there and simply acknowledge that, yes, hops has terrific flavors and helps prevent spoilage of the beer.

California sagebrush
(*Artemisia californica*)

But I've dealt with other plants, and their qualities and flavors, for a large part of my life, and by no means do I think hops should be the only herb that makes a beverage qualify as a beer. I prefer a pre-Reinheitsgebot definition, which might go like this:

BEER:
An alcoholic beverage made by the brewing and fermentation of bitter aromatic herbs, roots, bark, or other organic ingredients; sugar from malted grains or other sources such as molasses, unrefined brown sugar, tree sap (birch or maple syrup), and the like; yeast (wild or not); and water.

Essentially, if the beverage tastes like a beer, I will call it such, and I view hops as just one of the possible ingredients to create a beer. However, you're free to call nonhopped beers gruits or herbal beers if you wish—it's completely okay with me.

If you study and experiment with primitive and ancient beers, you'll find that an extensive set of plants was used instead of hops, and I'm pretty sure you can find one or more in your local environment.

Take the classifications below with a grain of salt, but over the years, this system has worked for me and helped me in creating beverages that exhibit beer-like flavors, qualities, and features.

Based on my research and experiments, the main herbs/ingredients used historically to brew gruits or herbal beers instead of hopped beers were as described below.

Category 1

This is what I call the Trinity, which refers to the main three ingredients I keep finding in the old recipes. You can use them separately (in mugwort beer, say, or yarrow beer) or mix them. They were used as the main flavor accents; all of them are quite bitter but highly aromatic. And of course, like hops, they have some very specific medicinal and psychotropic effects.

Yarrow (*Achillea millefolium*)
Bog myrtle or sweet gale (*Myrica gale*)
Mugwort (*Artemisia vulgaris*)

Most of these herbs are from Europe but can be grown or found in North America (which includes Canada). Yarrow is present in pretty much every state or province. Sweet gale exists as a native plant in the most northern states such as Oregon, Alaska, Vermont, and New York, and of course in Canada. Although non-native, mugwort is now found on the West Coast in Oregon, Washington, Alaska, and Canada and in every state on the East Coast. But we're lucky in California as we have our own native mugwort (*Artemisia douglasiana*), which can also be found in Oregon, Washington, Idaho, and Nevada. I think you can probably find it in Arizona, too. Mugwort is also common throughout Asia and North Africa.

As with regular beers made with hops, none of these three herbs should be consumed during pregnancy. Mugwort and sweet gale are known abortifacients.

Do you *have* to use of one those three herbs to make a gruit or herbal beer? No, you don't. Many flavorful and bitter beers have been made using other ingredients.

Category 2

The second category also includes bitter and aromatic herbs that can sometimes be brewed singly with other herbs and spices, but were used, often as flavor additives, with one or more of the Trinity herbs. Many adventurous home brewers also use them with hops.

> **Horehound** (*Marrubium vulgare*). I refer to horehound in my classes as the most bitter plant in the whole universe. It has been used medicinally as a remedy for respiratory ailments (horehound candies) and is a traditional herb used instead of hops in old beer recipes. Using it for brewing is an exercise in moderation unless you really love bitter beers. It's found in pretty much every region in North America (aside from North Dakota) and in parts of Canada. It's a non-native that came from Europe. I love using horehound with my bitter but highly aromatic local California sagebrush: I get fruity flavors that are reminiscent of some hop varieties.
>
> **Wild rosemary** (*Ledum palustre*). Found mostly in Alaska, Canada, and the northern and central parts of Europe, this plant (leaves and flowers) was used in traditional gruit recipes for its spicy and

bitter taste. Although it's an herb still used for medicinal purposes, it is presently not recommended for brewing due to its toxicity when consumed in larger amounts.

Wormwood (*Artemisia absinthium*). Native to temperate regions of Europe and North Africa, wormwood is presently naturalized in Canada and the northern United States. As the Latin name indicates, it's used to flavor absinthe. The plant is quite bitter but also highly aromatic. It's a bit of a controversial plant due to its thujone content; thujone is a toxic chemical that is harmful in fairly large quantities, though many people think the toxicity is overstated. (Thujone is also present in many other culinary herbs, such as sages.) Wormwood and absinthe are sometimes described as "psychotropic," but the evidence of hallucinogenic effects seems to have been grossly exaggerated in the literature related to absinthe. I use it sparingly in some of my wild beers.

Meadowsweet (*Filipendula ulmaria*). Also called mead-wort, this pleasant and aromatic herb has mostly been used as a flavoring in honey wine (mead) but also in old European beers. It is native to Europe but is presently naturalized in North America.

Heather (*Calluna vulgaris*) is found widely throughout Europe, and in some parts of Canada and the northeastern United States. It has medium bitterness and is quite flavorful (some think of it as similar to chamomile). The young, tender leaves and flowers are the parts most often used.

Dandelion (*Taraxacum officinale*). An extremely common edible plant in North America, its root and leaves are quite bitter and have been used as a bittering agent (mostly the root but sometimes the leaves). The flowers are used to make dandelion wine and are a good source of wild yeast.

Ground ivy (*Glechoma hederacea*). Native to Europe and southwestern Asia, this plant can be found in most parts of North America. It has been used as a preservative and flavoring in beers. The flavor is very interesting: fruity, mildly bitter, and peppery, too. It's one of my favorite plants for brewing.

Saint-John's-wort (*Hypericum perforatum*). A very medicinal plant (mostly used for depression and the healing of wounds and bruises), Saint-John's-wort was also used as a bittering agent in some European beers (particularly in Norway). It's native to Europe and Asia but now well established in North America as well. Presently the plant is known to adversely interact with a variety of pharmaceuticals, thus I haven't used it for brewing.

Common Herbs Traditionally Used for Brewing

Bog myrtle (*Myrica gale*)

Dandelion (*Taraxacum officinale*)

Horehound (*Marrubium vulgare*)

Yarrow (*Achillea millefolium*)

Category 3 (Spices, Flavorful Herbs, Fruits, and Berries)

This category comprises ingredients that people use to give their beer some interesting undertones: perhaps a bit of elderflower to impart a floral quality, orange or lemon peel for citric qualities, pepper or peppercorns for complexity and spicy notes, or even a mix of cloves, nutmeg, and cinnamon to give a pumpkin pie accent to your Thanksgiving brew.

As always, there are exceptions. Some beers, such as Belgian lambic beers featuring various fruits (raspberry, currant, and cherry, to name a few), use these ingredients as a main flavor.

I can't describe every plant, berry, and fruit listed in this book, but a lot of information is available online for each of these ingredients; a good search engine and the Latin name is all you need to get you started.

Allspice (*Pimenta dioica*)
Angelica
   (*Angelica archangelica*)
Anise (*Pimpinella anisum*)
Avens (*Geum urbanum*)
Basil (*Ocimum basilicum*)
Bitter orange peel (*Citrus* spp.)—
   used in some Belgian ales,
   with a citrus flavor profile
Burr chervil
   (*Anthriscus caucalis*)
Caraway (*Carum carvi*)
Cardamom (*Elettaria
   cardamomum*)
Cassia (*Cinnamomum cassia*)
Chamomile
   (*Chamaemelum nobile*)
Chamomile, wild, also
   known as pineapple weed
   (*Matricaria discoidea*)
Chile pepper (*Capsicum* spp.)
Cinnamon (*Cinnamomum* spp.)
Cloves (*Syzygium aromaticum*)
Cocoa (*Theobroma cacao*)
Coffee (*Coffea* spp.)
Coriander
   (*Coriandrum sativum*)

Costmary
   (*Tanacetum balsamita*)
Elderflower (*Sambucus* spp.)
Fennel (*Foeniculum vulgare*)
Fenugreek (*Trigonella
   foenum-graecum*)
Gentian root (*Gentiana lutea*)
Ginger (*Zingiber officinale*)
Ginseng (*Panax ginseng*)
Grains of paradise
   (*Aframomum melegueta*)
Hyssop (*Hyssopus officinalis*)
Lavender
   (*Lavandula angustifolia*)
Lemon peel (*Citrus* spp.)
Lemongrass
   (*Cymbopogon citratus*)
Licorice (*Glycyrrhiza glabra*)
Lime peel (*Citrus aurantiifolia*)
Nutmeg (*Myristica fragrans*)
Orange blossoms
   (*Citrus sinensis*)
Orange peel (*Citrus sinensis*)
Pepper, black (*Piper nigrum*)
Peppermint (*Mentha × piperita*)
Pink (*Dianthus armeria*)
Quassia (*Quassia excelsa*)

Rose hips (*Rosa* spp.)
Rosemary
  (*Rosmarinus officinalis*)
Saffron (*Crocus sativus*)
Sage (*Salvia* spp.)
Sassafras (*Sassafras albidum*)

Star anise (*Illicium verum*)
Vanilla (*Vanilla planifolia*)
Wild carrot (*Daucus carota*)
Woodruff (*Galium odoratum*)—
  usually added to an already
  fermented product

FRUITS AND BERRIES

Apricot	Date	Grapefruit	Peach
Blueberry	Elderberry	Juniper berries	Pomegranate
Cherry	Fig	Lemon	Strawberry
Currant	Grape	Lime	Et cetera . . .

Category 4

And now we go a bit extreme! Aside from the well-documented uses of category 1 and 2 herbs in herbal or hopped beers, other, more obscure plants, herbs, and organic ingredients have been used historically, often as bittering agents but sometimes also for flavors.

ROOTS

The best example of roots being used in brewing is the typical North American drink known as root beer. Root beer takes the tradition of brewing with what you can find in your local terroir to the extreme. It originated in the quests for survival, for rural medicine, and for local flavors. Although these days most recipes are nonalcoholic, older versions show that the drink followed the tradition of "small beers" with a low alcoholic content (usually around 2 percent). These were commonly used in Europe as daily drink instead of possibly polluted water.

A typical recipe for root beer included ingredients such as wintergreen leaf, gingerroot, sarsaparilla root, licorice root, sassafras root and bark, dandelion root, hops, birch bark, cherry bark, juniper berries, and cinnamon. Some recipes also called for yellow dock root, burdock root . . . as the name implies, it's a lot of roots!

Modern commercial root beers usually don't include sassafras root, which contains the aromatic oil safrole, banned by the FDA in 1960 because large doses can cause liver damage or cancer. The key term here is *large*: Many herbalists or root beer purists still use moderate amounts of sassafras root in traditional root beers, and I believe it's fine. You won't be drinking several gallons of this stuff daily (which would be equivalent to the dose of safrole that was given to lab rats in the studies).

Barks are another ingredient rarely mentioned in brewing discussions, but some of them have indeed been used traditionally. It makes sense: Barks can impart both flavors and medicinal benefits.

When I attended survival classes in the 1990s, for example, I learned that willow bark is the ancestor of aspirin, and if you've ever chewed on a small piece you know how bitter it is. And so I thought I was a genius when I came up with the idea of using of willow bark as a bittering agent in a medicinal beer . . . until a quick search online proved to me that my "new idea" was actually old news. Once again, I was just humbly rediscovering the forgotten past.

Thanks to the Internet and awesome search engines, you will find out that quite a few tree barks have been used either as bittering agents (such as alder or white oak bark) or for flavors (such as birch or cherry bark). Foraging is quite trendy these days, and quite a few small breweries are experimenting with the possibilities. In Ava, Illinois, Scratch Brewing Company makes a roasted hickory bark beer. For my part, I've used California sycamore bark for bitterness but also for the reddish color it imparts to the beer when boiled. I've also flavored one of my forest beers with toasted coast live oak bark.

I never collect bark from live trees. I usually take my willow and oak barks from freshly fallen trees, especially after a storm. California sycamore sheds its bark on a yearly basis; I collect this in early fall.

Once you start adventuring with barks, using other parts of the tree—wood, branches, leaves—is just another step. Wood chips, toasted or not, can be placed in your beer, soda, or mead to add interesting woodsy qualities. Lake Effect Brewing Company in Chicago uses oak chips from fallen trees to flavor their beer.

It's not a weird concept if you think about it. Many beers and wines are aged in oak barrels for the specific flavors the wood will impart over time. In fact, Anheuser-Busch uses beechwood chips in the brewing process of their flagship Budweiser brand.

What other wood chips, aside from oak, can be toasted and placed in your fermenting beer (or sodas or mead)? Try these:

Fig. Mild and nutty qualities
Manzanita. Mild and fruity, similar to applewood
Maple. Think maple syrup; the wood has caramel-like flavors
Mesquite. Strong but sweet; the flavor is a bit honey-like
Olive. Medium fruity flavors with a hint of mesquite
White ash. Light flavor, sweet and fruity
Yellow birch. Hints of root beer and caramel, medium flavor

I'm sure there are many other wood chips you could use, but they would require some thorough research. Juniper wood chips would probably work. I've seen online mentions in brewing forums of using cedar-, cherry-, and applewood chips. The fun is in researching what's available in your area and safe to use.

BRANCHES AND LEAVES

From barks and wood, let's move on to brewing with branches and leaves. Again, there's nothing new here; in fact the Finnish beer Sahti, one of the oldest beer styles in the world, uses whole juniper branches in the brewing process. I've had great success with adding pinyon pine branches to my cooled wort during the fermentation process. In Vermont we used yellow birch branches without the leaves for their sweet, root-beer-like flavor. One of my friends uses licorice chew sticks in his root beer recipe.

MUSHROOMS

Another rarely discussed ingredient is mushrooms, which can be used for flavors or medicinal benefits. Locally I use turkey tail mushrooms because of their slight bitterness but also their recognized health properties (they boost the immune system and have anti-cancer effects). We also have candy cap mushrooms, which are so packed with flavors that using them is an exercise in moderation; three to four of these mushrooms per gallon is usually enough.

It may not be mainstream, but the use of medicinal mushrooms such as chaga, shiitake, reishi, and others in beer-like beverages is well documented. Other mushrooms used for taste by various breweries include oyster mushrooms, chanterelles, boletes, morels, truffles, and probably many more.

I love mixing mushrooms when I make some of my "fall forest" beers. It's a great way to introduce earthiness, and the smells and tastes of mushrooms are strong reminders of autumn.

MOSSES AND LICHENS

These offer a whole new range of potential flavors begging for exploration. Lungwort (*Lobaria pulmonaria*), for instance, was used instead of hops by the monks of Ussokla monastery in Siberia. In Sweden sugar was extracted from various lichens in the late 1800s to be used for distillation. Irish moss (*Chondrus crispus*—it's actually a type of seaweed) has traditionally been used in beers to improve clarity.

NUTS

There are so many possibilities you can go nuts—literally! Perennial Artisan makes a black walnut ale, a peanut butter stout is brewed in Michigan, and if you do a bit of research online you'll find that hazelnuts, pecans, chestnuts, and many others have been used for flavoring. And of course, you have the famous myth (or is it?) mentioning an acorn beer made by the Pilgrims soon after they landed at Plymouth Harbor.

Category 5

A lot of other unusual ingredients are used to make various beers. I've put them into category 5 because they really don't fit within the flavor profile of a beer but often have a specific medicinal role. For example, a nettle or rhubarb beer, with its healthy content of vitamin C, would be a good remedy for scurvy. I think it's good to include a few of those traditional ingredients in this book—they could have a place in your experimental beers and other drinks, such as sodas or herbal meads.

By the way, although these ingredients are unusual, adventurous brewing companies have used them from time to time. For instance, the famous Cantillon brewery in Brussels, Belgium, has used rhubarb to create a sour Belgian beer.

Betony (*Stachys officinalis*)
Eyebright (*Euphrasia officinalis*)
Rhubarb (*Rheum rhabarbarum*)
Scurvy grass (*Cochlearia officinalis*)
Serviceberry (*Amelanchier alnifolia*)
Stinging nettles (*Urtica dioica*)
Watercress (*Nasturtium officinale*)
Wood sorrel (*Cochlearia* spp.)
Yellow or curly dock (*Rumex crispus*)

And if this isn't enough, some old beer and mead recipes will include ingredients such as propolis, bee pollen, royal jelly, violet leaves and flowers, parsley roots, nettle roots, mallow roots, fennel roots, strawberry roots, maidenhair fern, and so on.

So really . . . if it tastes good, fits within the flavor profile of the type of drink you're trying to make, and isn't actually *un*healthy, the sky is the limit.

If you're interested in delving deeper into unusual ingredients, I highly recommend the book *Sacred and Herbal Healing Beers* by Stephen Harrod Buhner. It's a large book—more than 500 pages—that includes a wonderful history of alcoholic beverages and countless recipes. It also discusses plants and other ingredients that were used for their powerful psychotropic effects, such as henbane, clary sage, and mandrake. These drinks, rooted in shamanic tradition, are beyond the scope of this book, but you can learn more about them in Buhner's volume.

What Do You Do with All These Categories?

Think of the above categories as rough guidelines about what ingredients you can use in your brewing. I do believe it's helpful to know them if you want to start experimenting. All are more or less traditional ingredients, and they may inspire you to create unusual sodas, meads, and other fermented drinks.

As you can see from the number of ingredients I've discussed (and I'm sure I could double or even triple that number if I wanted), there are no hard-and-fast rules for this; you can definitely mix and match. In my experience, the top categories (1 and 2 mostly) give you a fermented beverage with more "beer-like" qualities. When I made a forest beer in Vermont, though, I mostly used category 3 and 4 ingredients (white pine, blue spruce, dandelion, yellow birch branches, maple syrup, wild sassafras roots, and more), which was all I was able to find at the time. The end result was more like a spiced cider, though some people thought of it as a sort of beer.

Making Sodas, Herbal Meads, and Other Brews

In general, sodas are based more on fruity flavors than beers are.

Meads, like beers, come in a wide range of categories (around 20) depending on the type of ingredients used. For example, a *melomel* is a mead made using honey and fruit. A *metheglin* is a mead made with spices or herbs, and a *braggot* is a mead (or is it a beer?) made with both honey and malted grain.

I've given up on trying to fit my honey-fermented beverages into specific molds—because too often, as with my beermaking, I mix a wide variety of ingredients. So I simply call many of my meads "herbal meads" or just "mead." Because I use honey to make many of my sodas, technically I guess some of them could be called meads as well.

My point is, I really don't want you to think of the categories of ingredients used to make beer as applicable to beers only. These categories equally apply to making sodas or meads, or to flavoring wines.

Even bitter ingredients such as mugwort, hops, or yarrow make wonderful sodas if paired with the appropriate fruits. One of my favorite boozy sodas is made of prickly pears, mugwort, yarrow, and lemons. It's all in the amounts you use. It's actually quite fascinating: You can take the exact same ingredients and, by changing the proportions or the sugar source, create completely different drinks. For example, you can make a soda fruity by using prickly pears as the main ingredient, or make it taste like a beer by including molasses as the sugar source and mugwort or hops as the main flavoring. Heck, use honey as the main source of sugar and voilà! You're making mead.

Some Local Plants, Mushrooms, Fruits, and Berries I Use for Brewing

California juniper berries
(*Juniperus californica*)

California sagebrush
(*Artemisia californica*)

Elderflowers (*Sambucus mexicana*)

Manzanita berries (*Arctostaphylos glauca*, *A. glandulosa*, and *A. patula*)

Prickly pears (*Opuntia* spp.)

Sugar bush berries (*Rhus ovata*)

White fir (*Abies concolor*)

White sage (*Salvia apiana*)

Yerba santa (*Eriodictyon californicum*)

That's why I don't worry too much about trying to classify my drinks. It can be quite confusing, and I'd rather focus on research, flavors, and creativity. Beyond all the classifications and boundaries lies the exciting and often delicious world of "just having fun."

Brewing Your Own Local Terroir

From what you've read so far in this chapter, you would think that pretty much everything has been tried in brewing—yet that's far from true!

Brewing with what nature or your native plants provide you is where imagination flourishes, magic happens, and—unless you're living at the North Pole or in the Sahara Desert—with research and experimentation you will make unique and delicious beverages.

To give you an idea, I have so far identified more than 150 local wild ingredients I can brew with. These range from bitter herbs to berries, barks, mushrooms, aromatic plants, and roots. By combining these ingredients in different ways and amounts, I can create an infinite number of truly local, one-of-a-kind beers, sodas, meads, and wines.

Most of those ingredients cannot be found at the store. If you go to your local supermarket and ask for mugwort, yarrow, willow bark, or lemonade berries, you're probably going to get a weird look. If you totally rely on your local stores, you may find it hard to explore local flavors unless you work with your local native nurseries and plant some at home.

By creating your own "brewing garden," you can also help with the propagation of native plants, while creating habitat for butterflies and all kinds of pollinators, and benefiting the environment. Everybody wins.

I believe that researching and discovering your local flavors can go much deeper than just taste. A quest for local flavors is part of our hunter-gatherer DNA; it's a primal exploration of our surroundings and a discovery of what they have to offer. Over time it gives you a sense of place, of belonging.

For example, you can drink a wonderful sour beer and think it's a good representation of Belgium, or a delicious Bordeaux and taste that peculiar terroir through the intricate flavors of the wine. But if you are truly familiar with the region, your experience may go much deeper to evoke mental images of beautiful hillsides. You can nearly smell the fresh fragrance of the local flora and even the soil.

Sometimes I take a sip of one of my fermented drinks, close my eyes, and for a few seconds I'm back in the forest or the mountains. I can smell the local herbs, feel the wind and the joy of being there. It's a wonderful achievement when you can do all that with a drink!

Some Local Ingredients I Use for Brewing

Acorns (*Quercus* spp.)

Cactus pears (*Opuntia* spp.)

California juniper berries
(*Juniperus californica*)

California peppertree
(*Schinus molle*)

California sagebrush
(*Artemisia californica*)

California sycamore bark
(*Platanus racemosa*)

Grass (Poaceae family)

Lemonade berries
(*Rhus integrifolia*)

Local Mexican elderflowers
(*Sambucus mexicana*)

Local oak barks (*Quercus lobata*,
but many other oaks will
work as well)

Local willow bark (*Salix* spp.)

Pinyon pine (*Pinus monophylla*)

Rose hips (*Rosa californica*)

Sugar bush berries (*Rhus ovata*)

Toyon (*Heteromeles arbutifolia*)—
can only be collected on
private land

Tree leaves (willow, alder, maple)

White fir (*Abies concolor*)

Wild chamomile
(*Matricaria discoidea*)

Wild cherries (*Prunus ilicifolia*)

Wild currant (*Ribes* spp.)

Wild fennel (*Foeniculum vulgare*)

Woolly bluecurls
(*Trichostema lanatum*)

Yerba santa (*Eriodictyon
californicum*)

Hardcore Research

Fully exploring your terroir will require dedication. It's not something that will happen overnight. When I started learning about wild edible plants and berries in Arizona, California and Oregon, I probably attended over 400 classes and workshops given by naturalists, survivalists, botanists, local natives, wild food instructors, and food preservation experts over a period of four years.

Only after that period of learning the basic wild edibles as well as traditional and modern food preservation techniques did I start researching more deeply all the other plants and ingredients that weren't covered in those classes.

The biggest realization that hit me as I delved deeper into studying my surroundings is that the potential for new discoveries and creativity is really infinite. The more I learn about what nature can offer, the more I realize there is so much that I don't know. It's truly humbling.

On the plus side, with the help of regional experts or simply common knowledge, you can actually start your brewing exploration right away with basic wild (or not) edible fruits and other plants, such as blackberries and stinging nettles.

It's a valid place to start exploring local flavors, and as you keep studying you'll discover a whole universe of yummy concoctions to make.

More Local Plants, Mushrooms, Fruits, and Berries I Use for Brewing

Figs (*Ficus* spp.)—fruits and leaves

Forest grass (Poaceae family)

Forest leaves: willow and alder

Local mugwort (*Artemisia douglasiana*)

Pineapple weed (*Matricaria discoidea*)

Turkey tail mushrooms (*Trametes versicolor*)

Wild cherries (*Prunus ilicifolia*)

Wild currant berries (*Ribes* spp.)

Willow bark (*Salix* spp.)

I get excited when I discover new plants or ingredients that can be used for creating innovative and tasty brews, but there's a lot of hardcore research involved before you get to the actual brewing and tasting.

If you intend to serve your fermented concoctions to family, friends, or other people, it's paramount to make sure they're safe to consume. That can require a tremendous amount of investigation, which can last weeks, months, or sometimes years.

When I find a potential new brewing ingredient (plant, bark, berry, what have you), I go through a series of steps to make sure it can be used safely.

The first step is extremely important: proper identification. I can't stress that enough. Make sure you validate your own identification through the expertise of others. Now that we have the Internet, this is much easier than it was in the old days, and I strongly recommend that you become a member of various online plant identification groups. You can post photos of your find and get a confirmation that the plant is what you think it is. I go even further and make sure that local (live) experts agree, too, even if that means bringing the plant or berry to them.

From there, I start researching past culinary uses through books, but also online. Over the years I've collected a large number of books about wild edibles and ethnobotany. But the Internet is also a tremendous resource, and you can find databases there detailing the historical uses of plants by people of various regions or countries. For instance, there's the Native American Ethnobotany site (naeb.brit.org), which is a large database of foods, drugs, dyes, and fibers of Native American peoples. Or there's Plants for a Future (pfaf.org), a good searchable database of European plants.

Also, be sure to consult your local native plant nurseries. Very often they have local experts and botanists who can help you.

If you find that an ingredient has been used extensively in the past as food or medicine with no reported ill effects, it's a good start. As you do this type of research, you'll be amazed to see how much knowledge has been completely forgotten or lost.

But don't assume that because an ingredient has been used, it's completely safe to do so at the present time. Modern lab analysis sometimes shows that a plant traditionally used for brewing contains chemicals that aren't really healthy for you. For example, the young flowering tips of Scotch broom (*Cytisus scoparius*) were used for brewing in the past, before

---

Local ingredients foraged in Vermont to make a forest beer: yellow birch bark and branches (*Betula alleghaniensis*), blue spruce (*Picea pungens*), wild sassafras roots (*Sassafras albidum*), white pine (*Pinus strobus*), and yarrow (*Achillea millefolium*). Wild yeast comes from dandelion flowers, and maple syrup was used as the sugar source.

the introduction of hops, but the plant is now considered somewhat toxic due to alkaloid toxins (mostly found in the plant's "beans").

Even plants considered okay for brewing, such as mugwort, white fir, and others, should not be consumed if you are pregnant because of their abortive properties. Online resources such as WebMD, the National Capital Poison Center (www.poison.org), and countless others can help you in your research. I also make sure to check for potential allergies.

Be aware that some plants or spices are considered safe if consumed in small amounts, but can be toxic in larger quantities. A good example is nutmeg.

The idea is simple: Only use an ingredient when you are 100 percent *sure* that it's safe to do so.

Drying Your Herbs and Plants

Dehydrating wild plants properly is a bit of an art. As you experiment you may find temperatures and methods that work better for you than what I suggest. To be honest, when dealing with wild plants, I decide on the best way to dehydrate them based on their condition, the time of year, their aroma, and so forth. For example, young nettles in January are so fragile that I dry them carefully in paper bags, while the older ones will dry just fine in a dehydrator.

As a general rule I like to harvest plants for drying in the morning and ideally at the stage when they are about to start flowering. Many plants and herbs have volatile aromatic oils that evaporate to some degree after exposure to direct sunlight during the day. If necessary, gently rinse them in cold water before drying. I never dry herbs in direct sunlight. I air-dry them by hanging them in a paper bag in a well-ventilated area. Punch holes in the

Local mugwort drying in the shade.

bag to facilitate evaporation. Due to my location (California) and the normal high temperatures here, I don't always need to use paper bags (which protect plants from dust or insects)—the drying time is usually short.

To start your dehydrating experiments, divide your plants and herbs into three categories: fragile, medium, and tough. Each category calls for slightly different methods of drying.

Fragile Herbs/Plants

This category consists of plants that are tender and often have somewhat delicate aromatic properties, such as fennel or chervil.

Method: Pick fragile plants in the morning if you can, rinse if necessary, and pat them dry using a clean towel (a paper towel is fine) or shake to remove excess water. Air-dry them (in a paper bag with holes poked in it) in a well-ventilated area. Once the leaves are dry and brittle, place them in a tightly closed jar or another container. Store away from light.

If you live in a very humid area, air-drying may not be possible. In this case, use a dehydrator at a temperature set as high as 125°F (52°C) and store in an airtight container as soon as the leaves become dry and brittle.

Medium Herbs/Plants

This category consists of plants that are less delicate and retain their aromas or flavors easily, such as nettles, mints, and more.

Method: I forage the best specimens (usually in the shade) in the morning or early afternoon. Back home, I rinse them if necessary and pat dry using a clean towel or a shake to remove excess water.

Place in a dehydrator at a low setting, around 100°F (38°C), and store in an airtight container as soon as the leaves are dry and brittle. If you live in a very humid area, use a temperature as high as 125°F (52°C).

Tough Herbs/Plants

Tough plants retain their aromas or flavors easily. Examples include sages, mugwort, white fir, and pine needles.

Method: Rinse the plants, and shake off excess water. Place in a dehydrator at a medium setting, around 135°F (57°C), and store in an airtight container as soon as the leaves are dry and brittle.

In my experience, some "tough" herbs such as mugwort, eucalyptus, or bay leaves don't benefit from dehydrating rapidly and are much better air-dried on the stems.

CHAPTER 5

Methods of Brewing

I use three main brewing methods for herbal beers, sodas, and other fermented beverages. The ingredients themselves dictate which methods I use. There are no set rules as to any method you *must* use; it's all based on your preferences and your taste buds.

For example, I think that boiling mint isn't a good idea. Some of the wild mints I collect have very subtle and delicate flavors that can be destroyed or altered considerably during the boiling process. So if I make a mint soda, I would rather ferment it without boiling that herb. I simply place my (cleaned) mint in cold water, add some sugar (around ¾ cup/150 g white sugar or honey for ½ gallon/1.89 L) and yeast, and let it ferment for a couple of days before bottling.

The same thing would apply to pine or fir. I really don't like the taste of boiled pine at all, so if I make a pine soda I use *cold brewing*. However, if I make a beer that includes pine or fir but also ingredients such as hops or mugwort, I may decide to boil some of my herbs first, let the solution (wort) cool down, then pour it into a fermentation bucket or clay pot. Then I'll add my pinyon pine branches and yeast for a few days, or until the fermentation time called for in the original recipe is complete. It's a mix of two different methods—a *hot-and-cold brewing*.

If I make a horehound, hops, mugwort, or yarrow beer, it's much better to use *hot brewing*, because boiling is necessary to extract the flavors from those herbs. Ingredients such as mushrooms or barks require boiling as well.

You are completely free to have fun and experiment with each of these brewing methods, and over time you'll develop preferences based on what works for you. Here are some of my own preferences:

Cold-brewing ingredients collected during a morning hike: fig leaves, prickly pears, willow leaves, California sagebrush, water mint, and black sage. Wild yeast from prickly pears and raw honey were also used.

Ingredients and Plants I Use in Hot Brewing

Bitter (or not) plants with strong flavors such as mugwort, hops, yarrow, horehound, dandelion, white sage, yerba santa, grass, yellow birch, chamomile (and pineapple weed), and woolly bluecurls.

Berries such as elderberry, manzanita, currant, sumac, grapes, and blueberries.

Mushrooms (turkey tails, chaga, reishi); barks (willow, alder, oak); tree leaves (willow, alder); branches (yellow birch); and roots (licorice, sassafras, dandelion).

Ingredients and Plants I Use in Cold Brewing

Plants and other ingredients with subtle flavors, such as mints, basil, elderflower, white fir, pinyon pine, white pine, and fennel.

Berries and fruits such as cactus (prickly) pears, pineapple, pears, and juniper berries.

I use hot-and-cold brewing when I mix ingredients—say, in a mountain beer that includes yarrow, manzanita berries, and mugwort (better in hot brewing) but also pinyon pine and California juniper berries (better in cold brewing).

Note that some ingredients can benefit from a very short time in hot water. For example, elderflowers, thyme, tarragon, or pineapple weed are often placed in the brew when it's in the process of cooling down. You can also time your ingredients: It's extremely common for beermakers to place hops in the boiling wort at different times and for different durations to optimize flavors and aromas. With experience, you can do the same thing with your local herbs, managing their time of infusion in the boiling wort.

HOT BREWING

Hot brewing is used for plants or other ingredients that benefit from boiling to best extract their flavors. It's a very simple process that can be used to make sodas, wines, or beers. If you take a look at the fundamentals of brewing, you basically combine plants, sugar, and water and then add yeast. It's not much different from making tea.

What you need

Plants/ingredients as indicated in the recipe

Measuring cups

Scale

1 gallon (3.78 L) water—don't use tap water, which may contain chlorine

Brown sugar (or other type of sugar as specified in the recipe)

Large pot with lid

Sieve

Two funnels (one small, one big)

1-gallon (3.78 L) bottle or other container with airlock

Yeast (beer yeast for beer-type drinks; champagne, wine, or wild yeast for sodas)

Depending on the recipe, recycled soda bottles or swing-top bottles (at least seven 16-ounce/500 ml bottles for a 1-gallon batch)

Measuring spoon

A small quantity of white or brown sugar (used for priming at bottling time)

Procedure

1. Gather the ingredients you will use to make the beer, weighing all the ingredients as needed.

2. Place all the ingredients, along with the water and sugar, into a pot. Bring the solution to a boil and continue boiling for the length of time indicated in the recipe.

3. Place the pot into cold water to cool the solution to a lukewarm temperature of around 70°F (21°C). Keep the lid on to make sure airborne bacteria or insects such as flies don't infect your brew. Placing ice in the water will speed up the process. Change the water as necessary to cool the liquid faster.

4. Once the solution has cooled, pour the liquid through a sieve and funnel into your main fermenter (bottle or fermenting bucket).

5. Add the yeast. Note that you can also add the yeast into the cooled solution before you pour it into the fermenter.

6. Place a clean airlock into a rubber or plastic stopper on top of the fermenter. Fill the airlock with water up to the lines indicated. (Some people use vodka.)

7. Within 24 hours the fermentation should be active. You will see froth forming on top of the liquid and bubbles moving up in the airlock.

Step 3

8. Let the liquid ferment for the amount of time indicated in the recipe, then pour the contents, using a funnel, into swing-top bottles. Both the bottles and the funnel should be thoroughly cleaned before use. Wines such as elderberry or wild currant are usually fully fermented then placed into wine bottles and corked, as you'll see in wine recipes later on.

9. For sodas and beers, fill the bottles until the liquid reaches the beginning of the neck. For primitive beers, prime each bottle with ½ teaspoon (2 g) of regular white or brown sugar. Close the top and wait for a few weeks (usually 4 to 5 weeks), then enjoy! Sodas usually don't need additional sugar for carbonation.

Step 6

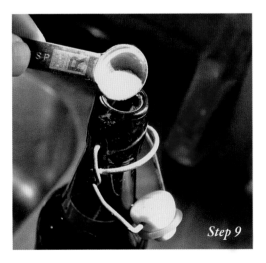

Step 9

COLD BREWING

Cold brewing is used for plants or other ingredients whose flavors can be altered, lost, or diminished if boiled. Some of my local mints are a good example. I also don't like the flavor of boiled California juniper berries or pinyon pine—to me, they taste too much like strong sap (think Pine-Sol cleaning solution)—but if you steep a couple of pinyon pine branches with needles cut at the top so the flavors can leach into your cold fermenting brew, you end up with some beautiful tangerine-like accents.

This type of brewing is really perfect for making sodas with fresh aromatic ingredients. You can create very tasty blends using spices, fruits, leaves, berries, and strong-tasting plants such as sages. It is the method I use to make my "hike in a bottle" sodas, whereby I gather ingredients while hiking and, once home, ferment what I've foraged to create beautiful drinks celebrating the flavors of a local environment. You could also do a "morning walk in the garden" soda.

What you need

Plants/ingredients as indicated in the recipe
Measuring cups
Scale
½–1 gallon (1.89–3.78 L) water—don't use tap water, which may contain chlorine
White sugar (or other type of sugar as specified in the recipe)
Yeast (usually champagne, wine, or wild yeast)
½-gallon to 1-gallon regular canning jar, European jar with wire bail and glass cap, or fermenting bucket with airlock (available at your local brewing supply store or online)
Sieve
Large funnel
Depending on the recipe, recycled soda bottles or swing-top bottles (at least seven 16-ounce/500 ml bottles for a 1-gallon batch)

Procedure

1. Gather the ingredients you will ferment and clean them thoroughly.
2. Place them into the jar or bucket you will use as your fermenting vessel and add water, sugar, and yeast as per the recipe.
3. Close the container. If you use a regular canning jar, you could place a paper towel on top and secure it with a rubber band. If you use a jar with a wire bail and glass cap, close it, but you will need to "burp" it later on (see page 94). The fermentation gases must be able to escape.
4. Three times a day, stir the solution for a minute or so using a clean spoon.
5. Within 24 hours, depending on the temperature, the fermentation should be active, and you will see froth/bubbles forming on top of the liquid.
6. Let the liquid ferment for the amount of time indicated in the recipe. You can enjoy the fermented drink as is, or you can strain and pour the contents, using a funnel, into recycled soda bottles or swing-top glass bottles. Both the bottles, strainer, and funnel should be thoroughly cleaned before use.

Step 1

Step 2

Step 3

Step 4

Step 5

Step 6

Cold brewing isn't the best method for long-term preservation. Due to the lack of boiling, you still have a lot of different bacteria and types of wild yeast at work that can alter the flavors over the long term. I've had a couple of aged drinks that turned into vinegar eventually . . . which wasn't a bad thing.

You can ferment wines for 2 to 4 weeks in the container or ferment them fully in a bottle (see the elderberry wine recipe on page 186), then place the wine into regular corked bottles.

Step 1

Step 2

Step 3

Step 4

Step 5

Step 6

HOT-AND-COLD BREWING

Hot-and-cold brewing is the method I use when I have ingredients that benefit from boiling to extract the flavors, such as roots, dehydrated berries, or aromatic plants like hops or yarrow, but as part of the recipe I also have plants whose flavors would be altered too much or simply lost in the boiling process, like mint, basil, or pine.

The process is simple. You divide the fermentation into two parts. First you boil the ingredients that will benefit from heat with your sugar source. Once this is done, cool your solution and place the liquid into a clean fermenting bucket fitted with an airlock or a jar with a wire bail and glass cap. Then add your more "sensitive" ingredients and the yeast to the wort. Let it ferment for the length of time called for in the recipe.

What you need

Plants/ingredients as indicated in the recipe

Measuring cups

Scale

1 gallon (3.78 L) water—don't use tap water, which may contain chlorine

Brown sugar (or other type of sugar as specified in the recipe)

Large pot with lid

Fermentation bucket with airlock or a jar with a wire bail and glass cap

Yeast (beer yeast for beer-type drinks; champagne, wine, or wild yeast for sodas)

Sieve

Two funnels (one small, one big)

Depending on the recipe, recycled soda bottles or swing-top bottles (at least seven 16-ounce/500 ml bottles for a 1-gallon batch)

Measuring spoon

A small quantity of white/brown sugar (for priming at bottling time)

Procedure

1. Gather, clean, weigh, and boil your hot-brew ingredients (including the water and sugar) as explained in steps 1 through 4 under "Hot Brewing" on page 81. Strain the solution into your clean fermentation bucket or jar.

2. Add the yeast and the other ingredients whose flavors will be extracted during the cold fermentation (for example, pinyon pine).

3. Place a clean airlock on top of your fermentation bucket or close the top if you used a jar with a wire bail and glass cap.

4. Within 24 hours the fermentation should be active. You will see froth forming on top of the liquid and bubbles moving up in the airlock. If you used a jar with a wire bail, you may need to burp it (see page 94). Note that you could also use a regular jar with a clean paper towel secured by a rubber band on top.

5. Let the liquid ferment for the amount of time indicated in the recipe, then pour the contents, using a funnel, into swing-top bottles. Both the bottles and funnel should be thoroughly cleaned before use.

6. Fill the bottles until the liquid reaches the bottom of the neck. For primitive beers, prime each bottle with ½ teaspoon (2 g) of regular white or brown sugar. Close the top and wait for a few weeks (usually 4 to 5 weeks), then enjoy! Sodas usually don't need additional sugar for carbonation.

Dry-Hopping

Even ingredients that often require boiling to extract the flavors can also benefit sometimes from cold fermentation. Very often, boiling does remove some of the subtle essences of herbs like hops or yarrow.

To remedy that, some brewers time the addition of hops into the boiling wort, adding a specific amount at the last minute or when the wort is removed from the heat and begins the process of cooling. It's called *late hopping*.

Other brewers go a little further and add dry hops to the vessel while the beer is fermenting (cold fermentation); these folks swear it's an excellent method to imbue the beer with the delicate aroma some hops are known for. If you're an herbalist, it's really a sort of alcoholic extraction occurring during the fermentation process.

Of course I had to experiment with the concept, and yes, I tend to agree about the process. Many of the strong bitter herbs I use instead of hops— such as yarrow, mugwort, or our local California sagebrush—have wonderful aromatics and essences that tend to get lost in boiling. But by adding a bit more when the solution is fermenting, you can elevate the flavors just a tad.

My main concern was the danger of bacteria infecting the brew and the possibility of a bad fermentation. However, if you think about it, strong herbs such as yarrow, some sages, mugwort, and my local California sagebrush all have antiseptic qualities, just like hops.

In my opinion some of the main bitter herbs don't benefit much from this technique. A good example is horehound. That mint is just packed with overwhelming bitterness, and subtle aromatics usually come from added herbs. But this is really the realm of opinion and a personal sense of flavor. I can see how someone could disagree with me.

As usual, the moral is that there are no set rules, and exceptions abound. As a brewer, you are free to mix and match techniques to achieve the ultimate flavors. It's true for beers, but also completely applicable to sodas, wines (primitive or not), and any other strange fermented concoctions.

By all means, experiment!

FERMENTING A COLD INFUSION

This is mostly a food preservation technique I use for some of my sodas. It's a slight variation on the usual cold-brewing technique.

The idea occurred to me after a private dinner we had a few years ago. For the occasion, I made 3 gallons (around 12 L) of a very elaborate infusion of locally foraged ingredients in cold water. The infusion was delicious, but at the end of the dinner I had 1 gallon (3.78 L) left. Instead of throwing it away, I decided to add yeast and additional sugar for another day, then strained and bottled the liquid. The end result was just as delicious.

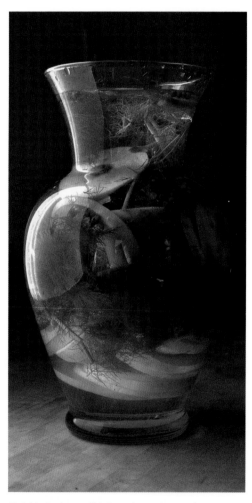

Cold infusion: wild fennel, apples, and pink peppercorn.

Cold infusion: oranges, peppermint, and elderflower.

The main difference between this and the usual cold-brewing technique (explained earlier) is that the ingredients are first infused in cold water (and in the fridge) for a day or two instead of being fermented from the start. This allows you to create and serve some delicious aguas frescas to guests, then save what's left over by turning it into a soda. The fermentation raises the acidity level of the liquid, which allows you to preserve it for a longer amount of time. Without fermentation, the liquid would probably go bad in a week or even less in the warm climate of Southern California.

Agua fresca (the term literally means "fresh water") is defined as a combination of ingredients—fruits, aromatic herbs, berries, barks, roots, cereals, flowers, and other similar ingredients—blended with water and sugar. It's very popular as a refreshing and nutritious nonalcoholic drink in South America, and probably in any hot country.

Cold infusion: elderflower, white fir, pinyon pine, manzanita, and California juniper berries.

The reason I use cold infusion, instead of placing the container in the sun as people do when making sun tea, is related to food safety. Yes, warm temperatures can speed up the infusion process, but I scrapped the idea because of the potential bacterial growth that can occur at higher temperatures. Dealing with fresh wild plants instead of dehydrated tea leaves, this issue is even more of a concern.

I experimented with placing my wild infusions overnight and up to a couple of days in the refrigerator, where the temperature is very low and bacterial growth isn't really a concern, and frankly I was amazed with the results. It worked beautifully with many aromatic plants, such as white fir, pine, yerba santa, mints, and my various wild berries, and to this day it's one of my favorite methods to create refreshing drinks. With the magic of fermentation, you can now take it to the next level, preserving your creation and brewing some amazing sodas.

Procedure

1. Forage your ingredients (from your garden, in the wild, or even at your local farmer's market). They can range from aromatic plants and berries to interesting ingredients such as pine needles, flowers, and so on.

   If you have a garden, you can create some wonderful flavor combinations using your planted herbs, including various mints, culinary sages, lemon balm, lavender, basil, thyme, rosemary, and of course local fruits.

2. Work out your mix. There are no real rules; you just need to experiment until you are happy with the flavors. As a forager, your infusions will also change greatly with the seasons and whatever is available. When I make my mountain infusion, I usually use 40 percent wild mint, 25 percent white fir or pinyon pine needles and branches, and the rest a mix of various ingredients, such as around 20 cracked California juniper berries, around 40 cracked manzanita berries, and a couple of sliced lemons (or lemonade berries). If I use yarrow or mugwort, I do so very sparingly—maybe a few leaves. Add sugar or honey to taste. Mints provide an excellent base on which to build flavor, and we have over seven found locally, each with different flavors.

   If you use white fir and pine, cut the needles first so it's easier to extract the flavors. You can also experiment with various woods; I've used California juniper wood chips with interesting results. In springtime I also crack small pinyon pine branches to access the incredibly flavorful sap.

3. Clean your container thoroughly—this is especially important if you plan to serve your infusions to others. After cleaning with soap and hot water, I often do a final rinse with very hot water.

4. Clean and place your garden or foraged ingredients and sugar source (if any) inside the container and add as much cold water as possible (don't use tap water, which often contains chlorine and other chemicals). Close the container. I often recycle old vases or glass containers, placing plastic wrap on top. If you try this, make sure the containers are lead-free. Food-grade plastic buckets are great, too.

5. Place the container in the refrigerator for at least 12 hours and up to 48 hours. When the mixture is done, I usually strain it into a new and beautiful container with a spigot before serving it to my guests, placing some of the ingredients I used for the infusion into the serving container, purely for aesthetics. When you're done serving the guests—or if you decided from the start to do a cold infusion just to make soda—simply add sugar and water (1½ to 2 cups/ 300 to 400 g per gallon) to the liquid and ferment at room temperature for a day. You should see some bubbling occurring, indicating the fermentation is active.

   The liquid is then strained and placed into recycled plastic soda bottles or swing-top bottles. Usually 8 hours is enough time for bottle conditioning. You can check the carbonation by feeling the pressure with the plastic soda bottles or by slightly opening one of your swing-top glass bottles (see "Wild Carbonation in Bottles" on page 99). When you're satisfied with the level of carbonation, place the bottles in the fridge; this will slow the fermentation. I like to drink my soda within a week.

   As you'll see later, herbal meads are similar. The main differences are that the yeast and honey or sugar are added from the start, and the liquid isn't placed in the fridge because the fermentation process raises the acidity level (pH) very quickly, making it safe to drink.

Hygiene

Hygiene is important, but you don't have to be obsessive about it. Realize that beers and wines have been made for a very long time, and some periods in history, such as the Middle Ages, were not known for their superior hygiene.

I've fermented all kinds of liquids for many, many years and presently don't purchase any sanitizers from my brewing supply store. That said, I clean everything with very hot water and (sometimes) regular dish soap before using it. I've never had one of my beers or sodas go bad . . . so far.

From time to time, I like to clean and sterilize my swing-top bottles by placing them in a dilute bleach solution for 5 minutes. Use 1 tablespoon (20 g) of bleach per gallon, and rinse the bottles several times afterward with hot tap water.

Fermentation Vessels

When I started brewing, I only used 1-, 3-, or 5-gallon (3.78–18.93 L) bottles or carboys, but because I do a lot of ferments in small quantities, as well as various experiments with local flavors, these days I use all kinds of containers including regular Mason canning jars and even recycled vases. (Make sure to test vases for lead, though—you can purchase lead testing kits online.)

I still use bottles for brewing beers, but jars are much better when you're dealing with plants, large fruits, berries, roots, and so on. With jars, you need to make sure that fermentation gases can escape but bacteria and critters from the outside cannot get in. If I use a regular Mason jar, I make sure that I don't screw the lid so tight that extreme pressure can build up inside. The same principle would apply to European-style jars that have a wire bail and glass cap; if gases can't escape, you'll need to "burp" your jar several times a day during active fermentation.

A lot of the European-style jars I purchase in the United States are mostly decorative and not very useful for home canning, so they actually work nicely for fermentation because the seal is often weak, letting the gases escape but not allowing anything from outside to get in. Still, if you get a new jar, you will need to make sure this is true. In a case of very active fermentation, usually in the first week, you need to supervise your jar. Aside from creating a mess, an exploding jar or bottle can be quite dangerous.

Placing a clean paper towel or regular towel on top of a fermenting vessel works very well, but it's not as effective a barrier against bacteria. On the plus side, I've never had a fermentation go bad using that technique. Note that sometimes the fermentation bubbles will push the ingredients up inside the jars, and liquid can spill over a bit. To avoid this, make sure you leave some room between the liquid and the top of your container.

"Burping"

Some people use a technique called "burping" whereby the pressure inside Mason jars (or similar containers) can be released two or three times a day for the first 3 to 5 days by unscrewing the lid; do this less often as the fermentation become less active. The technique is often used for lacto-fermentation when making sauerkraut, kimchi, and similar ferments. My own preference is to use the method of not screwing the lid too tight in the first place, so that fermentation gases can still escape. The only time I have used burping was with a European wire bail jar that had a very tight seal.

Fermentation Containers

European-style jars with a wire bail and glass cap. I "burp" the jar a few times each day if necessary.

Mason canning jars—usually quart or 1/2-gallon size. Screw the lids "loosely tight," but not tight enough to stop the fermentation gases from escaping. Shake around three times a day.

A clay pot made by my talented friend Melissa Brown Bidermann.

Recycled vase and clean towel. Test for lead before using vases or old crocks for fermentation.

Carbonation

Presently, all beers and many other drinks we buy at the store are carbonated. We've come to take it for granted, and carbonation is now seen as a quality and aroma enhancer.

There are two types of carbonation, forced and natural.

Nowadays, carbonated drinks have pretty much nothing to do with actual natural carbonation. If you buy a regular soda at the store, it's flavored sugary water into which carbon dioxide has been mixed under high pressure. It's the same with most commercial and even craft beers: Once the fermentation is complete and the beer has been placed into a keg, CO_2 is forced into the keg to create the carbonation.

In the natural process, carbonation is a normal by-product of fermentation within an enclosed container capable of sustaining a decent amount of internal pressure. Fermentation releases CO_2, and under pressure it's absorbed back into the liquid.

It's likely that some ancient beers or wines could have achieved some level of carbonation when stored in closed oak barrels or amphorae, but the consensus seems to be that carbonated drinks came with the invention of stronger glass bottles that were capable of sustaining the internal pressure from trapped CO_2.

Nevertheless, and just for fun, a couple of years ago I experimented with creating carbonated brews in primitive containers such as gourds and clay pots. I managed to get some interesting results by adding weight on top of the containers and using crude seals, sometimes made of grass. So it's possible that our ancestors made carbonated drinks, but I'm not sure if it was something they were even looking for.

Many very young and actively fermenting drinks have a slight carbonation, but it's mostly due to the activity of the yeast converting the sugar to alcohol rather than carbonation through pressure.

Unless I use them for cooking, most of the beers I make are carbonated. It's exactly the same process as making soda. As per the actual recipes you'll find in this book, the fermenting liquid is poured into swing-top bottles or recycled soda bottles (for sodas). A bit of sugar is added to feed the yeast inside the bottle (this is known as *priming*), and the bottles are closed. The liquid continues to ferment for a while, releasing CO_2 that, due to the pressure inside, is dissolved back into the brew.

Primitive carbonation achieved by placing a heavy rock on top of a clay container to create pressure and a tight-ish seal made from dry grass. *Pottery made by Melissa Brown Bidermann.*

A primitive carbonation system for wild beers: The pressure from the bow keeps the top closed. A light carbonation was achieved. *Photo courtesy of Mia Wasilevich.*

It's simple stuff, but if there's too much sugar left, you may end up with too much pressure inside. Hence it's a good idea in the beginning to follow recipes until you get a bit more experience.

Another method that's used to achieve carbonation with conventional grain-based beers (see "Civilized Carbonation in Bottles," page 100) could probably be adapted to primitive beers using malt as their main sugar source. The technique is to let the fermentation go until it's done, usually a couple of weeks, then add a bit of sugar or honey into the fermented liquid before bottling it. Because some yeast is still alive, the liquid will ferment a bit more in the bottle and create the carbonation. Using that method, you can control much more precisely how much pressure you get inside and thus use conventional capped beer bottles.

To be honest, I haven't used the latest method with many of my primitive beers, because I mostly use sugar and not malt. But if you want to use wild or garden ingredients, malt as the sugar source, and regular capped bottles, keep reading and give it a try.

Wild Carbonation in Bottles: Feel the Pressure

Because many of our primitive beers and sodas are bottled while the fermentation is still active, you will sometimes deal with some interesting variations in terms of fermentation activity, pressure, and carbonation. Some herbs, such as nettles or grass, can create a beautiful and very active fermentation, while some sages will create a more finicky and slow one. In the beginning it's a good idea to keep a journal so you can improve your beverages as you go along.

The yeast you use is also very important. For example, commercial champagne yeast is very reliable, but once you start dealing with wild yeast starters, you're entering the fermentation wilderness. From empirical experience, some strains seem to be more active than others. Temperature is also part of the equation; in winter fermentation is slower due to the colder temperatures.

So what do you do? The answer is simple: Observe!

As you gain experience, you'll realize there are no set rules. Guidelines and recipes are great in the beginning, but as you become a master at your craft, and especially when you're using wild or unusual ingredients, you'll realize that observation is more important. If you have a very active fermentation for a soda, you may not have to ferment it for 24 hours. Again: There are no set rules. The key is to observe. You want

to see a nice active fermentation, a healthy bubbling, and it should smell good. Then you can bottle it.

Let's assume you've achieved a nice bubbly concoction and bottled it in a recycled soda bottle or a swing-top glass bottle. Now what?

A very active fermentation inside a closed bottle can create a lot of pressure, and I've opened my share of bottles only to have liquid spurt out aggressively. When you're just getting started making sodas or other carbonated drinks, you may want to use recycled plastic soda bottles, because you can gauge the carbonation by pressing their sides with your fingers. When I was a beginner, I used a regular Coke bottle (filled with Coke) to compare the pressures. When I was satisfied that my own wild soda bottles had achieved a similar pressure, I would place them in the fridge. The cool temperature would considerably slow the fermentation process.

Additionally, take a look at the bottom of a plastic soda bottle: It's designed to pop out in case of too much pressure. If one day you see that the bottom of a plastic soda bottle has expanded and the bottle is starting to look like a balloon, you'd better find a safe space (outdoors) and carefully open the bottle, because the liquid inside will expel with a lot of force.

With a swing-top glass bottle, you can't tell how much pressure is inside unless you actually open it. My current technique, if the initial fermentation was very active, is to take one bottle from a specific batch and check the carbonation after 8 hours or so for sodas by holding and pushing with my left hand on top while releasing the side handles slowly with my right hand. Then, if I'm satisfied with the carbonation, I just place the bottles in the fridge. I go through the same routine before serving a batch to guests.

Sometimes I've miscalculated and achieved too much pressure. In that case I simply hold and release the top slowly a few times. Bit by bit, this gets rid of the excess pressure without losing any of the precious liquid. If I don't have enough pressure when the bottles come out of the fridge, I let them come back up to room temperature for a few hours. The fermentation process will restart, and I eventually get the right amount of carbonation.

Civilized Carbonation in Bottles: Priming, the Scientific Way

Here's the easy method: Ferment the liquid until all signs of fermentation activity stop. (Don't age it.) Some yeast will still be alive. Add a specific amount of sugar to your fermented liquid and mix it in well—I usually shake my 1-gallon (3.78 L) bottle for 30 seconds or so—then transfer the liquid into bottles. The specific amount of sugar you use will create a precise amount of carbonation.

What I like about this system is that if you follow a recipe exactly and complete a full fermentation before priming, you can obtain more consistent pressure, carbonation, and even flavors. Still, speaking personally, absolute consistency is not always a quality that I'm looking for. Many of my brews are a reflection of the local terroir and seasons. As such the flavors and even carbonation will change from batch to batch due to numerous variables. It's also rare when I make the same beer over and over; more often I add interesting seasonal ingredients such as different berries or aromatic herbs. Most of my wild beers are also drunk within 3 to 4 weeks and often have some residual sugars in them, which I like. In fact, I sometimes age some beers for just a couple of weeks if I intend to use them for cooking.

If you do a full fermentation before priming and use commercial yeast, you can usually achieve a higher alcohol level and a beer that is more dry or sour, because most of the sugar has been consumed and converted into alcohol by the yeast. If you want a sweeter beer, you can add an unfermentable sugar, such as lactose.

Somehow the flavors created using full fermentation work nicely with grain-based beers but, in my experience, not always with sugar- or molasses-based beers. So you'll need to experiment a bit, especially if you start mixing sugar sources such as malt and molasses. Remember the Neanderthal rule—1 pound (454 g) of sugar per gallon will give you around 5 percent alcohol—and also the fact that many wild yeast strains die at a concentration of around 5 percent alcohol. If you do a full fermentation with wild yeast, you may not have any yeast left to create carbonation, and the beverage may still be quite sugary (it has happened to me).

But by all means, play with this priming method a bit and decide for yourself. For sugar- or molasses-based beers, I prefer the noncivilized method and follow the usual "wild" recipes (a few days of fermentation, then bottle with minimal priming).

Even President George Washington, in his small beer (low-alcohol beer) recipe using molasses, recommended bottling the beer after a week of fermentation, seemingly without priming:

> Take a large Siffer [Sifter] full of Bran Hops to your Taste.–Boil these 3 hours then strain out 30 Gall[ons] into a cooler put in 3 Gall[ons] Molasses while the Beer is Scalding hot or rather draw the Molasses into the cooler & St[r]ain the Beer on it while boiling Hot. Let this stand till it is little more than Blood warm then put in a quart of Yea[s]t if the Weather is very Cold cover it with a Blank[et] & let it Work in the Cooler 24 hours

then put it into the Cask–leave the bung open till it is almost don[e] Working–Bottle it that day Week it was Brewed.

For grain-based beer, the flavors of fermented malted grains seem to go quite well with a full fermentation first, so I use this method if I use grains as my sugar source and go full fermentation then exact priming. The cool thing about exact priming is that you can calculate the right amount of sugar you need to get a specific carbonation pressure. In the brewing world, that pressure is called *CO$_2$ volume*. Once you can do that, you can use regular beer bottles and cap your beer without the risk of explosion.

How much priming sugar and CO$_2$ volume do you need? It's really your choice. Beer connoisseurs know that some types of beers are best enjoyed with a specific amount of carbonation. I like to use the same pressure as basic commercial beers such as Budweiser, Stella Artois, or Corona, which is around 2.5.

To get a CO$_2$ volume (carbonation pressure) of 2.5, here's how much priming sugar or syrup you will need per gallon (3.78 L) with your bottle stored at room temperature (72°F/22°C):

White or brown cane sugar: 0.9 ounce (26 g) per gallon
Corn syrup: 1.2 ounces (34 g) per gallon
Honey: 1.2 ounces (34 g) per gallon
Maple syrup: 1.1 ounces (30 g) per gallon
Molasses: 1.2 ounces (34 g) per gallon

This comparison can help you decide if this method suits you when making brown-sugar- or molasses-based beers. In my humble opinion, because of flavors, I still *much* prefer to bottle live beers (ones that are still fermenting) with minimal priming into swing-top bottles, but with regular hops and grains-based beers, including brewing with malt extract, I think civilized priming is the way to go.

Sodas are supposed to be sweet, so don't do a full fermentation for sodas unless you're looking for a "sour" quality—mostly if you're dealing with wild yeast.

MUGWORT–LEMON BEER

If you want to compare a full fermentation with the usual method of bottling while the fermentation is still active, try the regular Mugwort–Lemon Beer recipe.

1 gallon (3.78 L) water

0.3 ounce (9 g) dried mugwort

3 lemons

1¼ pounds (567 g) brown sugar (this should give you a fully fermented beer of around 6–7 percent alcohol)

Yeast (commercial or wild yeast starter)

Procedure

1. Boil the water, mugwort, lemons, and sugar for 30 minutes. Cool the solution, add the yeast, and strain into a 1-gallon (3.78 L) bottle.
2. Ferment until complete.
3. Prime by adding 0.9 ounce (26 g) brown sugar to the fermented liquid, then bottle it. Wait 2 to 3 weeks and enjoy.

HOPPY BROWN SUGAR BEER

Or you might want to make a beer with hops and brown sugar to compare with a regular commercial beer:

1 gallon (3.78 L) water

0.25–0.4 ounce (7–11 g) hops of your choice (I choose Amarillo pellets, because I like mild beers, and use the lower amount)

0.8 pound (12.8 ounces/363 g) brown sugar (this should give you a fully fermented beer of around 4 percent alcohol)

Yeast (commercial or wild yeast starter)

Procedure

1. Boil the water, hops, and sugar for 30 minutes, adding half of the hops at the beginning and the other half after 20 minutes.
2. Cool the solution, add the yeast, and strain into a 1-gallon (3.78 L) bottle.
3. Ferment until complete.
4. Prime by adding 0.9 ounce (26 g) brown sugar to the fermented liquid, then bottle it. Wait 2 to 3 weeks and enjoy.

The Zen of Fermenting

One more tip before we get to the actual recipes. It mainly applies to cold-brewed sodas, wines, or similar beverages. Here it is: When you're making a specific fermented beverage, you can simply follow a recipe or a set procedure and get good results, but over the years I've taught classes in which we all tasted the difference between some of my wines and others that were made from the exact same ingredients by students. Often the comments were: "I like Pascal's version much better."

The difference isn't really that I'm a better brewer or have some sort of secret skill. Granted, experience plays a role, but the major factors are *care* and—dare I say—*relationship*. A ferment is very much like a living organism, and to some degree I establish a caring relationship with it. I always have several concoctions fermenting, and they're an intricate part of my life. I take care of them and they take care of me later on.

My day usually starts at around 5:30 AM. The first thing I do is check on my babies, see how they're doing, make sure ingredients remain under the brine for my lacto-fermentations, watch my airlocks to see how active my various alcoholic fermentation projects are, take a clean spoon and taste things if necessary. Each fermentation project is different, and you learn so much by watching their evolution. A good example was my sweet cactus pear "wine" laced with a bit of mugwort and yarrow. I started with these ingredients in ½ gallon (1.89 L) of water:

> 12 large cactus pears, sliced in half
> 10 crushed California juniper berries
> 3 yarrow flowers
> ½ pound (227 g) brown sugar (to start with)
> 0.15 ounce (4 g) dried mugwort
> 1 lemon, sliced

The prickly pears and juniper berries supply the brew's wild yeast. You can ferment these ingredients for a week using the cold fermentation method, but after a while, when the yeast has eaten all the sugar, it becomes quite sour and unpleasant. Typically, you'd add sugar at the end of the fermentation, but too often the flavors are affected when the liquid goes sour. You may end up with a so-so drink. Another approach would be to add much more sugar in the beginning so there is leftover sweetness once the wild yeast has died (usually at 5 percent alcohol), but wild yeast can be finicky and may decide to give up at 4 or 6 percent. You might end up with either a dry or a slightly sugary wine.

So instead of following an exact routine or recipe, I just check my wine daily and make sure I keep it at the optimal flavor. Usually I don't touch it for the first 5 to 6 days, as the fermentation is ramping up. At that point I take a clean spoon and taste a small amount. If it's going a bit sour already, I add some sugar. If it's a bit too bitter, I may remove some of the mugwort.

I do this every day or two and usually shake the jar two or three times a day. By taking care of the fermentation, adding sugar as I go along, I can keep a ferment going for weeks until it has achieved the amount of alcohol I was looking for. During the whole process, I keep the flavors exactly the way I want them to be by not allowing the wine to go sour or the herbs to become overwhelming. (You can always remove or add more herbs.) The wine can be drunk at any time during the fermentation process, by the way; some people like it with very little alcohol. When you're happy with it, just place it in the fridge, which will slow down the fermentation considerably, and drink within a week or so. You can also make this type of wine with plums, grapes, pears, cherries, and more.

For my prickly pear wine, I ended up using around 1½ pounds (680 g) of sugar (50 percent brown sugar, 50 percent white). I opted for making a sweet drink as I went along, simply because it balanced well with the bitterness of mugwort/yarrow and tasted much better that way.

If I can do this, anyone can. All it takes is a bit of care and observation. Become one with your brew!

CHAPTER 6

Sugar, Molasses, and Syrup-Based Beers

L et's celebrate how easy it is to make some delicious fermented drinks. In this chapter we'll keep things super simple and, as much as possible, use what you already have in your kitchen or garden or what you can find locally in nature (for instance, dandelions or nettles). The recipes here are straightforward and will get you started on the right path.

As I explained at the beginning of this book, I make most of my home-made "beers" using various sugar sources, such as brown sugar, molasses, or even tree saps (maple or birch syrup). This pretty much guarantees that the beverage I end up with tastes like a cross between beer and cider. The recipes that use bitter ingredients such as mugwort, horehound, yarrow, and (especially) hops have a flavor profile much closer to beer.

These basic recipes are not written in stone; you can play with them, add more ingredients, and experiment. For example, you can substitute culinary sage for white sage, or use the mugwort-cranberry recipe with your local sour berries. It's all good!

As you read the recipes, let your creative spirit go wild and think of how you can adapt them to use what's available to you locally. Take a walk in your garden or a hike through nature, smelling or tasting the plants and letting nature inspire you. The idea is to have fun and create something that will taste good in the end. Of course, if you do any foraging, make sure that you can identify whatever plants you find with 100 percent certainty.

Don't dismiss culinary herbs or spices. My local Middle Eastern market has interesting fermented beverages made from rosemary, thyme, and tarra-gon, to name a few herbs. So why not give them a try and see what happens? Adding a bit of rosemary and orange zest, or tarragon and lemon, may end up tasting awesome!

Just imagine what you might do with the herbal tea recipe! You can brew countless kinds of herbal "tea-beers" using commercial blends; one Internet site listed more than 100. I'm all about flavor, but if you're into natural healing, take a look at the herbal blends available to you locally; at my own store I saw blends developed for common ailments such as cold and flu, sleeping disorders, stress, and much more. And don't forget that you can also create your own herbal blends to work with.

You don't always have to brew 1 gallon (3.78 L) if you're experimenting with herbal tea-beers; just adapt the recipe to the amount you're making. Most of the time, I just brew ½ gallon (1.89 L) or even less if I'm trying something new. You can use recycled wine bottles or jars.

You'll notice I use a lot of lemons or "sour/lemony" ingredients in some of my brews; it's just a personal preference. You can try the recipes with or without these additives, or replace them with other interesting flavors such as orange or grapefruit peel, pineapple rind, and so on.

MUGWORT-CRANBERRY (OR LEMON) BEER

Mugwort (*Artemisia vulgaris*), a culinary and medicinal herb, can be found growing throughout much of North America, Europe, and Asia. On the West Coast and in parts of Mexico, we have a native mugwort (*A. douglasiana*) that can be used instead of the common species. The plant was used as an aromatic and bittering agent in old traditional European beer recipes before hops became widespread. The plant has also digestive, antibacterial, and antifungal qualities.

1 gallon (3.78 L) water
0.3 ounce (9 g) dried mugwort leaves
1¼ pounds (567g) dark brown sugar
3–4 cups (around 1 L) cracked cranberries, or 3 large lemons
Yeast (beer yeast or wild yeast)

Procedure (see Hot Brewing on page 81)

1. Mix the water, mugwort, and brown sugar in a large pot. Using a glass or stone, crush the cranberries and place them in the pot. If you're using lemons, squeeze out the juice with your hands then throw them in the pot. Bring the solution to a boil; let it boil for 20 to 30 minutes.

2. Remove the pot from the heat and place it in cold water. Cool to 70°F (21°C), then add the yeast (wild or commercial). One 5-gram (0.176-ounce) packet of commercial yeast is usually enough for 5 gallons (18.9 L), so you don't need to use the whole packet. If I'm using a wild yeast starter, I usually use a bit more than ½ cup (120 ml) of liquid.

3. Strain the brew into the fermenter. Position the airlock or cover your fermenter with a paper towel or cheesecloth. Let the brew ferment for 10 days. Start counting when the fermentation is active (this may take 2 to 3 days with a wild yeast starter).

4. Siphon into 16-ounce (500 ml) swing-top beer bottles (you'll need seven bottles) and prime each one with ½ teaspoon (2 g) white or brown sugar for carbonation. Close the bottles and store in a place that's not too hot. The beer will be ready to drink in 3 to 4 weeks.

California mugwort (*Artemisia douglasiana*)

HOREHOUND BEER

Horehound (*Marrubium vulgare*) is a medicinal plant (used for colds and flu) from the mint family that is native to Europe and Asia but now found in most of North and South America. It's probably the most bitter plant I know and was used traditionally as a bittering agent for beer. To this day horehound candies are still made as a cough remedy. If you're brewing with this herb, remember that a little goes a long way. Most recipes I've found use a greater quantity of horehound than I do, possibly because they use it fresh, not dried.

1 gallon (3.78 L) water

0.07 ounce (2 g) dried horehound (you can use 0.1 ounce/3 g or more if you like a more bitter beer)

0.1 ounce (3 g) dried California sagebrush (optional)

1¼ pounds (567 g) dark brown sugar

2–3 lemons, or 0.3 ounce (9 g) lemongrass

Yeast (beer yeast or wild yeast)

Procedure (see Hot Brewing on page 81)

1. Mix the water, horehound, sagebrush, and brown sugar in a large pot. Cut and squeeze the lemons into the pot. If using lemongrass, place it in the water now. Bring the solution to a boil; let it boil for 20 to 30 minutes.

2. Remove the pot from the heat and place it in cold water. Cool to 70°F (21°C), then add the yeast (wild or commercial). One packet of commercial yeast is usually enough for 5 gallons (18.9 L), so you don't need to use the whole packet. If I'm using a wild yeast starter, I usually use a bit more than ½ cup (120 ml) of liquid.

3. Strain the brew into the fermenter. Position the airlock or cover your fermenter with a paper towel or cheesecloth. Let the brew ferment for 10 days. Start counting when the fermentation is active (this may take 2 to 3 days with a wild yeast starter).

4. Siphon into 16-ounce (500 ml) swing-top beer bottles (you'll need seven bottles) and prime each one with ½ teaspoon (2 g) white or brown sugar for carbonation. Close the bottles and store in a place that's not too hot. The beer will be ready to drink in 3 to 4 weeks.

YARROW BEER

Yarrow (*Achillea millefolium*) is a very aromatic but bitter medicinal herb native to Asia, Europe, and North America. Historically it's been used topically, for its healing properties on wounds. Yarrow was also one of the main bittering agents used in Old World beers before the introduction of hops. From experience, probably due to location and soil conditions, the flavors of yarrow can range from mild to overwhelmingly bitter, so you may need to adjust the recipe a bit. Thanks to its strong aromatic qualities, yarrow is also making a comeback as a culinary herb.

1 gallon (3.78 L) water
0.45 ounce (around 12 g) yarrow flower heads and leaves
¾ pound (340 g) brown sugar + 6 ounces (180 ml) molasses (or substitute 1¼ pounds/567 g dark brown sugar, or 1 pound/454 g dry malt extract)
Yeast (beer yeast or wild yeast)

Procedure (see Hot Brewing on page 81)

1. Mix the water, yarrow, brown sugar, and molasses in a large pot. Bring the solution to a boil; let it boil for 30 minutes. For more flavor, you can add some of the yarrow 5 minutes before the end of the boiling time.
2. Remove the pot from the heat and place it in cold water. Cool to 70°F (21°C), then add the yeast (wild or commercial). One packet of commercial yeast is usually enough for 5 gallons (18.9 L), so you don't need to use the whole packet. If I'm using a wild yeast starter, I usually use a bit more than ½ cup (120 ml) of liquid.
3. Strain the brew into the fermenter. Position the airlock or cover your fermenter with a paper towel or cheesecloth. Let the brew ferment for 10 days. Start counting when the fermentation is active (this may take 2 to 3 days with a wild yeast starter).
4. Siphon into 16-ounce (500 ml) swing-top beer bottles (you'll need seven bottles) and prime each one with ½ teaspoon (2 g) white or brown sugar for carbonation. Close the bottles and store in a place that's not too hot. The beer will be ready to drink in 3 to 4 weeks.

Using Dry or Liquid Malt Extract

If you like to brew with malt, you can make all these basic recipes using dry or liquid malt extract instead of brown sugar, molasses, maple syrup, or other sugar source. For 1-gallon (3.78 L) recipes, I usually use 1 pound (454 g) dry malt extract or 1¼ pounds (567 g) liquid malt extract. You should end up with a beer having around 4.5 percent ABV.

The procedure is identical to that used for sugar-based beers. Doing a full fermentation (not just 10 days) then priming works well with malt extract. See "Civilized Carbonation in Bottles" on page 100.

NETTLE BEER

I sometimes enjoy going back to my Celtic roots and making some medicinal herbal beer. Aside from making a somewhat pleasant drink, stinging nettles (*Urtica dioica*) are a great source of vitamin C and apparently help alleviate rheumatic pain. It's a Belgian dream—a beer with a lot of health benefits. Different brewers use different methods for this recipe. What follows is what I do.

1 gallon (3.78 L) water
1–1½ pounds (454–680 g) fresh nettles
¾–1¼ pounds (340–567 g) brown sugar
   or molasses
½ ounce (14 g) gingerroot (cut finely or grated)
3 lemons, or 1 cup (250 ml) sumac berries
2 ounces (56 g) dandelion leaves (optional)
1 ounce (28 g) cream of tartar
Yeast (beer yeast or wild yeast)

Procedure

1. Bring the water to a boil and add the cleaned nettles. Boil for 20 to 30 minutes, then add the brown sugar, ginger, lemons (juice them first, then throw them in the brew as well), dandelion leaves (if using), and cream of tartar. Boil for another 5 minutes.

2. Remove the pot from the heat and place it (with the lid on) in your sink filled with cold water. Change the cold water in the sink two or three times until your beer is lukewarm (around 70°F/21°C).

3. Strain into your fermenter (bottle, pot, or whatever you're using), add the yeast (wild or commercial), and place an airlock (or clean towel) on top.

4. Ferment for 3 to 4 days for ¾ pound (340 g) sugar or 7 to 8 days for 1¼ pounds (567 g) sugar (counting from when the fermentation is active, usually 8 to 12 hours after adding the yeast), then bottle. I don't use any priming sugar. The fermentation is quite active after bottling so I like to use a recycled plastic soda bottle to monitor for any excess carbonation and release it if necessary by opening the top slowly. Most people drink nettle beer young, after a week or so; it's not meant to be aged, as the flavors will be altered, and not in a good way.

DANDELION BEER

Dandelions (*Taraxacum officinale*) can be found pretty much anywhere in the world. Like nettle beer, dandelion beer is deeply rooted (no pun intended) in the tradition of brewing medicinal herbal beers. It's mostly used as a tonic (the plant is a rich source of beta-carotene and vitamin C) but also offers health benefits for liver disorders, urinary disorders, and diabetes. The flowers are used to make wine and are a good source of wild yeast. I like the beer more for its health benefits than its taste (not my favorite).

1 gallon (3.78 L) water
½ pound (227 g) fresh dandelion greens
½ ounce (14 g) chopped dried dandelion roots
   (often available from natural food stores)
1 pound (454 g) brown sugar
½ ounce (14 g) gingerroot (cut finely or grated)
1 ounce (28 g) chopped fresh lemongrass,
   or 3 lemons (optional)
1 ounce (28 g) cream of tartar
Yeast (beer yeast or wild yeast)

Procedure

1. Bring the water to a boil and add the fresh greens and dried roots. Boil for 20 to 30 minutes, then add the brown sugar, the ginger, the optional lemons (juice them first, then throw them in the brew as well), and the cream of tartar. Boil for another 5 minutes.
2. Remove the pot from the heat and place it (with the lid on) in cold water. Change the cold water two or three times until your beer is lukewarm (around 70°F/21°C).
3. Strain into your fermenter (bottle, pot, or whatever you're using), add the yeast (wild or commercial), and place an airlock (or clean towel) on top.
4. Ferment for 7 days. Start counting when the fermentation is active (this may take 2 to 3 days with a wild yeast starter), then bottle. I don't use any priming sugar. The fermentation is active, so I like to use a recycled plastic soda bottle to monitor for any excess carbonation and release it if necessary by opening the top slowly. The beer is meant to be drunk young, usually after 7 to 10 days,

MAPLE BEER

When I did some research on old and often forgotten homemade beers, probably the simplest I found was Maple Beer. How simple was it? Here are the ingredients:

2½ pounds (1 quart) or 1.13 kg (1 L)
    maple syrup
1 gallon (3.78 L) water
Yeast

Procedure

The method described was extremely simple: Bring the syrup-water solution to a boil, cool it down, place into the fermenter, and add the yeast. Do a full fermentation (see "Civilized Carbonation in Bottles" on page 100), then bottle, prime, and cap. It was drunk within a week.

The final beer is quite alcoholic, and if you use wild yeast, the amount of syrup is a bit excessive. I would use around 1¾ pounds (680–794 g) or 1.6 cups (0.4 L) syrup; otherwise the beer might end up quite sugary.

In his book *Sacred and Herbal Healing Beers*, author Stephen Harrod Buhner has a recipe using 3 pounds (0.25 gallon) or 1.36 kg (1 L) of maple syrup per gallon and another interesting recipe using 2 gallons (7.5 L) of fresh sap and yeast. Three pounds (1.36 kg) of sugar could be fine so long as you use a commercial beer yeast, which allows for a higher percentage of alcohol.

After making it and tasting it, I found this maple beer pretty bland, to be honest. I love using maple syrup, but it's much better with hops, mugwort, or other bitter herbs.

GINGER BEER

Another simple but interesting beer you can try at home is Ginger Beer.

10 ounces (284 g) light brown sugar
   (some people use white sugar)
1 gallon (3.78 L) water
Around 2 ounces (56 g) gingerroot, cut thin
2 lemons
Commercial or wild yeast

Procedure

Place the sugar in the water and bring it to a boil. Remove from the heat and add the ginger and lemon juice (throw the lemons in as well). Cover the pot or pour everything into a new container with a lid on. When the solution has cooled to 70°F (21°C), add the yeast. Let everything steep for 48 hours (stir a couple of times daily), then strain into recycled soda bottles or swing-top bottles. Check the pressure (see page 99) after 10 hours or so and place in the fridge if appropriate. It should be ready to drink in 3 to 4 days.

Some similar recipes just strain the liquid into a gallon bottle with an airlock, do a complete fermentation, then bottle. The end result is really a cross between a beer and a soda, but it's quite delicious.

By the way, you can add 0.15 ounce (4 g) mugwort or yarrow to this recipe for additional local wild flavors and a more bitter "beer-like" quality, but feel free to experiment with any other local berries or plants you may have.

CULINARY BEERS:
ROSEMARY, BAY, AND OTHER HERBS

You can use the following recipe and method to make beer from any of the various herbs in your garden—tarragon, thyme, basil, and so on. These beers were originally intended for medicinal use: Rosemary is said to ease inflammation and muscle pain and helps the immune system, while bay was traditionally used for colds and flu, muscle pain, breaking fevers, and to aid digestion.

I'm sure that in the old days, home brewers were making all kinds of interesting medicinal concoctions, and many of those recipes have long been lost to time. But I think it's a good idea to research such archaic brews, experiment with them, and see if they have a place in the kitchen—if not for drinking, then maybe for cooking or making vinegars. You'll find similar brews in Stephen Buhner's book, Andy Hamilton's book *Booze for Free* (I have not read this, but I've come across a few of his brewing experiments online), and various brewing forums where herbal ingredients are also used in more traditional hops beers as added flavors.

Like the nettles or dandelion beers, these herbal beers are not meant to be aged for very long and usually won't age well at all. I recently had a sip of my nettle beer, which had been bottled for 4 weeks . . . bad idea! But when consumed fresh they often can be enjoyable or interesting, especially from a culinary perspective.

Think like a cook or a chef and investigate which spices or herbs are used in various dishes. Rosemary is often used with lamb or poultry; I could see some chicken, lamb, or goat cooked in rosemary-bay beer. If you're vegan, it could be great for cooking carrots. Tarragon and fennel are often used with fish recipes, but a beer or soda could also make a nice sorbet or granita. Lavender can be used with pork, but is also used in pastries (lavender beer muffins . . . anyone?). Thyme has a wide variety of culinary and medicinal uses.

Take this basic recipe and play with it a bit, maybe adding some citrus, sour grass, or orange zest. Then try using the same method with other culinary herbs, see what comes out, and tweak if necessary.

1 gallon (3.78 L) water—don't use tap water, which may contain chlorine

1 pound (454 g) brown sugar or molasses

2 large sprigs fresh rosemary (you can use dried if that's what you have)

3–4 regular bay leaves

Commercial or wild yeast

Procedure

1. Mix the water, brown sugar, rosemary, and bay leaves in a large pot. Bring the solution to a boil for 20 minutes.

2. Place the pot into a pan of cold water; cool to 70°F (21°C), then add the yeast. When I'm using a wild yeast starter, I usually add a bit more than ½ cup (120 ml) of liquid, up to ¾ cup (180 ml).

3. Strain the brew into the fermenter. Position the airlock or cover your fermenter with a paper towel or cheesecloth. Let the brew ferment for 7 days. Start counting when the fermentation is active (this may take 2 to 3 days with a wild yeast starter).

4. Siphon into 16-ounce (500 ml) swing-top beer bottles (you'll need seven bottles) and prime each one with ½ teaspoon (2 g) white or brown sugar for carbonation. Close the bottles and store in a place that's not too hot. Drink within a week or two. For cooking or culinary uses, you can also freeze the beer (but not in the bottle).

Lamb stewed in rosemary-bay beer and California sagebrush stems.

The Wildcrafting Brewer

HERBAL TEA BAG BEERS

This goes into the category of "Why not?" and is probably the simplest recipe in this book. But seriously, you can make some very nice brews by using commercial organic herbal blends or exploring the fermenting possibilities with your favorite herbal tea to see how it translates into a beer or soda. You'll be amazed! I've made a terrific root beer by purchasing a commercial root beer blend. It's not as much fun as growing or foraging the ingredients yourself, but sometimes you don't have a choice. There are literally hundreds of herbal teas and blends for purchase at your local store and online.

This is super easy to brew.

1 gallon (3.78 L) water
1¼ pounds (567 g) brown sugar (you can also experiment with around 14 fluid ounces/415 ml molasses or maple syrup)
Your favorite tea bags (see manufacturer's instructions for quantities)
Commercial beer yeast or wild yeast starter

Procedure

1. Place the water and sugar into a pot and bring it to a boil. Remove from the heat and steep the tea bags as per the manufacturer's instructions. Usually 5 to 10 minutes is called for, but I go by flavor and often steep the bags longer.
2. Remove the bags and place the pot (with the lid on) in a sink filled with cold water. Change the cold water two or three times until your brew is lukewarm (around 70°F/21°C).
3. Strain into your fermenter (bottle, pot, or whatever vessel you're using). Add the yeast (wild or commercial) and place an airlock (or clean towel) on top. When I'm using a wild yeast starter, I usually use a bit more than ½ cup (120 ml) of liquid.
4. Ferment for 10 days. Start counting when the fermentation is active (this may take 2 to 3 days with a wild yeast starter).
5. Siphon into 16-ounce (500 ml) swing-top beer bottles (you'll need seven bottles) and prime each one with ½ teaspoon (2 g) white or brown sugar for carbonation. Close the bottles and store in a place that's not too hot. The beer will be ready to drink in 3 to 4 weeks.

   You can also use less sugar and ferment for a shorter time, as in the root beer recipe on page 134.

CUSTOM HERBAL BLEND BEERS

Once you're hooked on the awesome brews you can make with commercial herbal tea blends, you'll start making your own blends using what your garden or nature has to offer. The potential for creative fun is huge. You can enjoy the results simply as herbal teas or ferment them into sodas or beers. You'll need to experiment a bit with the blends and amounts—sometimes the fermentation will alter the flavors, but that's all part of exploring your local terroir. The accompanying photo shows one of my favorite "chaparral blends," composed of woolly bluecurls (*Trichostema lanatum*—buy plants from a nursery), local manzanita and sumac berries, a type of wild chamomile called pineapple weed, and a tad of yarrow.

The method is simple. Basically you're making sugary herbal tea, then adding sugar or honey.

1 gallon (3.78 L) water

1¼ pounds (567 g) brown sugar (you can also experiment with around 14 fluid ounces/415 ml molasses or maple syrup)

½–1 cup (120–250 ml) of your favorite homemade dried herbal blend (the quantity really depends on the type of ingredients you use; you may need to experiment a bit)

Commercial beer yeast or wild yeast starter (you could experiment with champagne yeast or wine yeast in some blends, mostly if you include berries, dried fruits, and aromatic, nonbitter plants)

Procedure

1. Place the water and sugar into a pot and bring it to a boil. Remove from the heat and steep your herbal blend. Then *taste*! Judging by the flavors, add more of the blend if necessary; stop the steeping when you're happy. Take note of steeping time and quantity for the future. Strain the liquid into a clean container.

2. Place the container (preferably with a cover on) in a sink filled with cold water. Change the cold water two or three times until your brew is lukewarm (around 70°F/21°C). Pour into your fermenter (bottle, pot, or whatever you're using), add the yeast (wild or commercial), and place an airlock (or clean towel) on top. When I'm using a wild yeast starter, I usually use a bit more than ½ cup (120 ml) of liquid.

3. Ferment for 10 days. Start counting when the fermentation is active (this may take 2 to 3 days with a wild yeast starter).

4. Siphon into 16-ounce (473 ml) swing-top beer bottles (you'll need seven bottles) and prime each one with ½ teaspoon (2 g) white or brown sugar for carbonation. Close the bottles and store in a place that's not too hot. The beer will be ready to drink in 3 to 4 weeks.

   You can also use less sugar and ferment for less time, as in the root beer recipe on page 134.

Simple Sugar-Based Beers Using Hops

Hops are the flowers of the hop plant (*Humulus lupulus*). What's used to make beer in North America is mostly European imports, although native hops do grow here and you can also find hops growing wild (as a non-native) in many states with moist, temperate climates such as Oregon, Washington, Idaho, and New York. Of course, it's easy to purchase hops online or at your local brewing supply store.

Unlike many other plants used to make beer-like beverages, such as horehound or mugwort, you must be a bit careful and selective when using hops. Why? Because in the last few hundred years, and pretty much solely due to its use in beermaking, the hop plant has been one of the most modified, altered, enhanced, and worshipped plants on earth.

I won't lie to you: I'm not an expert on the subject. There are whole books dedicated to hops, detailing everything from how to grow them to the attributes of different varieties. So this discussion may oversimplify things, but hopefully it'll work for now. If you want more in-depth information, it's easy to find.

Hops are usually sold either as pellets or as dry flowers. Some are extremely bitter and others very aromatic, hence the following problem: A recipe can't just call for 0.2 ounce (6 g) of dry hops flowers and call it a day. Much depends on the type of hops. That 0.2 ounce may leave your beverage slightly bitter but highly aromatic—or give you a beer that's too bitter for your taste.

There are three main categories of hops: bittering hops, aromatic hops, and hops having both qualities. The bitterness of the hops comes from their alpha acid resin. This resin is usually insoluble in water unless it is boiled. The longer the boil, the more bitterness you can extract.

There is an actual scale (IBU, or international bittering units) of bitterness based on the amount of alpha acid content. It's helpful, but understand that bitterness is also a highly personal perception. What might be mildly bitter for one person can be too much for someone else. You'll have to experiment a bit with hops—maybe start with my recipe (which may or may not be too bitter for you) and work from there. If you're really into hops, by all means purchase books on the subject and join the ranks of the enlightened!

Presently, hops' alpha acid content ranges from 1 to 17 percent. The higher the percentage, the more bitter the brew. The Hoppy Maply Beer recipe on page 129 calls for a type of hops that's on the mild side but still has nice aromatics. Here are some of the most commonly used hop varieties, with their alpha acid range:

Low-bitterness hops. Cascade (5–7%), Crystal (3.5–6%), Fuggle (5%), Golding (4–6%), Saaz (4–5%), Willamette (4–6%)

Medium-bitterness hops. Amarillo (7–11%), Centennial (9–11%), Magnum (12%)

High-bitterness hops. Chinook (12–14%), CTZ (14–17%), Nugget (11.5–14%), Simcoe (12–14%)

So far my favorite dual-purpose hops (with a nice balance of bitter and aromatic qualities) are Willamette, Amarillo, and Cascade, but I haven't tried them all. If you like my recipe but want to use another type of hops, think of it this way: You can theorically double the bitterness of your beer by using the same quantity of hops but choosing a type that has twice the alpha acid percentage. Or you can make your brew less bitter by choosing hops with a lower alpha acid percentage.

Brewers often mix various types of hops and adjust the boiling time to achieve the flavors they like. There is much more to know about hops; I hope this information offers a brief introduction.

HOPPY MAPLY BEER

1 gallon (3.78 L) water

½ pound (¾ cup/180 ml) maple syrup

10 ounces (282 g) brown sugar

0.15–0.20 ounce (around 4.2–6.0 g) dry hop flowers (I used Cascade hops with 5 percent IBU)

Yeast (beer yeast or wild yeast)

Procedure (see Hot Brewing on page 81)

1. Mix the water, maple syrup, and brown sugar in a large pot. Bring the solution to a boil and immediately add 0.05 ounce (1.4 g) of hop flowers. Let it boil for 30 minutes, then add another 0.05 ounce; boil for 20 minutes more and add the remaining hops. Boil for a final 10 minutes (for a 60-minute boil in all).

2. Remove the pot from the heat and place it in cold water. Cool to 70°F (21°C), then add the yeast (wild or commercial). If I'm using a wild yeast starter, I usually use ½ to ¾ cup (120–180 ml) of liquid.

3. Strain the brew into the fermenter. Position the airlock or cover your fermenter with a paper towel or cheesecloth. Let the brew ferment for 10 days. Start counting when the fermentation is active (this may take 2 to 3 days with a wild yeast starter).

4. Siphon into 16-ounce (473 ml) swing-top beer bottles (you'll need seven bottles) and prime each one with ½ teaspoon (2 g) white or brown sugar for carbonation. Close the bottles and store in a place that's not too hot. The beer will be ready to drink in 3 to 4 weeks.

Notes: As a rough guideline, if you use hop pellets instead of dry flowers, reduce the amount a tiny bit (around 5 to 10 percent) to achieve the same level of bitterness. So instead of using 0.15 ounce (4.2 g) of dry flowers, you would use around 0.13 ounce (3.7 g) of hop pellets.

Also, you can make this recipe without the maple syrup by substituting 1¼ pounds (567 g) brown sugar or 1 pound (454 g) dry malt extract, then doing a full fermentation and proper priming to achieve the right carbonation (see "Civilized Carbonation in Bottles" on page 100).

Root Beer

Root beer is truly an American drink. If you study its history, it is thought to have been based on Native American beverages that were consumed for flavor and/or medicinal uses.

The original recipes were probably adopted quickly by settlers and, in the European tradition of making low-alcohol beverages instead of drinking water, turned into mildly alcoholic beverages.

I think root beer is an acquired taste. I had my first root beer shortly after I immigrated to the United States in my mid-twenties. To be honest, it wasn't my favorite beverage. My first impression was that it tasted like medicine. But I can see how it could be comforting or considered delicious if you grew up with the flavors.

In the early 19th century, the drink was marketed as a healthy and nutritional tonic, often sold in stores as a syrup to which you added water. Although the recipes varied, the main ingredient in many traditional root beers is sassafras roots. In the early 1900s Dr. Swett's root beer listed its ingredients as "the juices of such beneficial berries, barks, roots, and herbs as checkerberry, sweet birch, sassafras, sarsaparilla, spikenard, juniper, wintergreen, ginger and hops."

Presently what you're purchasing at the store is a far cry from the original root beer. It's similar to other commercial sodas: carbonated water, corn syrup, caramel color, various preservatives, and natural and artificial flavors.

Modern root beers also don't use sassafras roots and bark. Safrole, the oil that gave root beer one of its main original flavors, has been banned from commercial uses by the FDA since 1960 due to the potential for liver damage and various types of cancers with long-term exposure. More recently the US National Toxicology Program (https://ntp.niehs.nih.gov) was milder in its assessment. They concluded that safrole is "reasonably" considered to be a human carcinogen based on studies in experimental animals, but they also mentioned that no human studies had been done to evaluate the relationship between exposure and human cancer. So who knows?

Some traditionalists think that the health benefits of

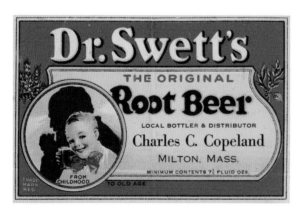

sassafras outweigh the dangers, and these people are quick to point out the large quantities used in studies, the fact that risks are only seen with long-term exposure, and the lack of proven consequences of exposure for humans. Many continue to make traditional root beer for their own personal consumption, as it was done for many generations. Sassafras roots for tea are still being sold as a dietary supplement, which is not regulated by the FDA.

My friend, renowned forager and hunter Hank Shaw, did some personal research and stated the following on his website: "Just know that there are many times more 'known carcinogens' in a bottle of beer than there are in any homemade sassafras product you might make. By one calculation, you'd need to drink 24 gallons of sassafras root beer a day for an extended time to get the amount of safrole fed to those rats."

Personally, I came to the same conclusion as Hank. But you should investigate the subject and make your own educated decision.

The recipes for root beer vary immensely. Some basic ingredients are always used to establish the main flavor profile, but after that it's all about making your own creative concoction. There are few hard-and-fast rules: Root beer may be alcoholic or nonalcoholic, carbonated or still, and may even contain caffeine.

Based on my own research, what follows are some of the ingredients used in modern and traditional recipes.

MAIN FLAVOR PROFILE INGREDIENTS (THESE MAKE IT TASTE LIKE WHAT PEOPLE THINK OF AS "ROOT BEER")

Anise (*Pimpinella anisum*)
Fennel roots and seeds
    (*Foeniculum vulgare*)
Hoja santa leaves (*Piper auritum*)
Licorice root (*Glycyrrhiza glabra*)
Sarsaparilla root (*Smilax regelii*)
Sassafras roots and bark
    (*Sassafras albidum*)
Wintergreen berries and leaves
    (*Gaultheria procumbens*)

INGREDIENTS THAT IMBUE SPECIFIC CHARACTERISTICS (THESE LET YOU MAKE IT YOUR OWN IN TERMS OF FLAVORS)

Balsam fir (*Abies balsamea*)
Black cherry (*Prunus serotina*)
Burdock roots (*Arctium lappa*)
Curly dock roots
    (*Rumex crispus*)
Dandelion roots
    (*Taraxacum officinale*)
Ginger stem and rhizome
    (*Zingiber officinale*)
Hops (*Humulus lupulus*)
Mugwort (*Artemisia vulgaris*)
Spruce (*Picea* spp.)—black,
    blue, red, and perhaps other
    types of spruce

Accents (spices and herbs)

Allspice (*Pimenta dioica*)

Astragalus root
(*Astragalus membranaceus*)

Black peppercorns (*Piper nigrum*)

Cassia bark (*Cinnamomum cassia*)

Chocolate (*Theobroma cacao*)

Cinnamon bark
(*Cinnamomum verum*)

Clove (*Syzygium aromaticum*)

Coffee (*Coffea* spp.)

Fenugreek (*Trigonella foenum-graecum*)

Juniper berries (*Juniperus communis* or other edible types of juniper berries)

Mint (*Mentha* spp.)

Nutmeg (*Myristica fragrans*)

Pinyon pine (*Pinus edulis*)

Star anise (*Illicium verum*)

Vanilla (*Vanilla planifolia*)

Possible Sugar Sources

Birch syrup

Brown sugar

Corn syrup (modern ingredient)

Honey

Maple syrup

Molasses

By all means, do some research online; you will find an incredible number of recipes to inspire you. If you want to try a very basic recipe for 1 gallon, you could use the following:

1 gallon (3.78 L) water

2 tablespoons (12 g) sassafras root bark

1 tablespoon (5 g) sarsaparilla root, or ½ teaspoon wintergreen leaf

1 teaspoon (6 g) dried licorice root

Small piece (¾–1 inch/1.9–2.5 cm) gingerroot

1½ cups (300 g) brown sugar

Yeast (wild or champagne/wine yeast; beer yeast would work, too)

SOUTHERN CALIFORNIA ROOT BEER

Of course, I had to add some ingredients to the basic recipe to give it a unique Southern California flavor.

1 gallon (3.78 L) water
2 tablespoons (12 g) sassafras root bark
1 tablespoon (5 g) sarsaparilla root
1 teaspoon (4 g) wild fennel seeds
Small piece gingerroot (¾–1 inch/ 1.9–2.5 cm)
0.1 ounce (3 g) mugwort (*Artemisia douglasiana*)
1 tablespoon (5 g) dandelion root
¼ cup (60 ml) lime juice
1 cup (200 g) brown sugar
½ cup (120 ml) molasses
10–15 California juniper berries (cracked)
2 small pinyon pine branches with needles (cut the tips)
½–¾ cup (120–180 ml) wild yeast (ginger bug)

Procedure

1. Bring the water to a boil and add all your ingredients except the juniper berries, pinyon pine, and yeast. Reduce the heat to a slow simmer and simmer for 20 to 30 minutes.

2. Remove the pot from the heat and place it (with the lid on) in a sink filled with cold water. Change the cold water two or three times until your beer is lukewarm (around 70°F/21°C).

3. Strain into your fermenter (fermenting bucket or large jar), add the yeast (wild or commercial), juniper berries, and pinyon pine branches, then place an airlock (or clean towel) on top. If I'm using a wild yeast starter, I usually use ½ to ¾ cup (120–180 ml) of liquid.

4. Ferment for 3 to 5 days. Start counting when the fermentation is active (this may take 2 to 3 days with a wild yeast starter). More days will make the brew more alcoholic and less sugary. Strain and transfer to bottles (such as recycled soda bottles); allow these to sit at room temperature for 12 hours, then place them in the refrigerator. Open slowly before serving to make sure the pressure is not excessive (see "Wild Carbonation in Bottles" on page 99).

Brewing Local

These days I hear the expression "Eating Local" all the time. It's a wonderful idea, but a bit misleading. It's really a nice marketing slogan designed to push consumers to purchase locally grown fruits and vegetables, but all too often these products have nothing to do with the actual terroir.

This is especially obvious here in Southern California, where so many products are grown. I would venture to say that 90 to 95 percent of what I find at the local farmer's market are edible plants that didn't grow here naturally in the first place. In all my time spent outside, I've never seen tomatoes, potatoes, leeks, grapefruit, lemons, bananas, cabbage, and so on growing untended in the wild environment.

Conversely, I don't find a lot of really local ingredients sold at the farmer's market, such as acorns, pinyon pine nuts, local berries, and aromatic plants. From my perspective, it's a bit bizarre.

If you think about it, a lot of the original land and truly local products have been erased and replaced by farmed non-native fruits and vegetables. From a culinary perspective, which includes creating drinks, I believe that by experimenting again with what the original environment provides we can start raising awareness of local edible plants, fruits, berries, or vegetables, and start using and enjoying them.

Who knows: If wild foods and going *truly* local become more popular, people will grow more native gardens at home instead of lawns, and maybe some farmlands will restore some of their original flora, so that native plants can grow once again in a sustainable way to the benefit of all—humans and animals alike.

Don't get me wrong. I purchase most of my fruits and vegetables from local organic farms, and I'm all into helping our local small farmers. I just want to point out that "Eating Local" should go beyond introduced fruits or vegetables. It could only be a good thing to re-establish a balance between introduced and original ingredients. I think our local farmers would be thrilled to plant more native fruits, berries, and herbs if there were a demand for them.

What Is the Advantage of Using Truly Local Ingredients?

When I was teaching at Sterling College in Vermont, we created a truly local primitive beer using ingredients from their forest (white pine, yellow birch, sassafras, blue spruce, et cetera), plus their own maple syrup—still produced as it was two centuries ago—and wild yeast from local birch trees and wildflowers.

We then used the beer to cook a type of stew incorporating local products such as apples, blueberries, and rabbit. Not only was the final dish incredibly tasty, but to this day you could serve it to me blindfolded and I would recognize the flavors of Vermont. You could not re-create that dish anywhere else.

Here in Southern California we have interesting local plants that are now being investigated for their culinary potential, and these flavors, too, cannot be reproduced anywhere else. For example, a local company called Ventura Spirits advertises on their website Wilder Gin, which is distilled with wild-harvested, native California botanicals including sagebrush, purple sage, California bay, and yerba santa.

By using what the actual terroir can offer, not only can you can experience brand-new flavors, but you can also create drinks that define your place. I don't care where you live: Unless it's the North Pole or the Sahara Desert, the potential for creativity is immense. It's just a matter of thorough research and experimentation. So let's create, and find ways to help the environment in the process.

CALIFORNIA SAGEBRUSH BEER

California sagebrush (*Artemisia californica*) is found in west-central and southwestern California and northern areas of Mexico. Locally, the plant has been used as a spice and also for medicine (cough, cold, and pain relief, and more). As an artemisia, the plant is related to wormwood and mugwort, both used as hops substitutes. It's the only "sagebrush" I know locally that has culinary uses. Like many hops substitutes, the plant is highly aromatic and bitter. I often use California sagebrush as an added aromatic in my horehound beer.

1 gallon (3.78 L) water

0.2 ounce (6 g) dried California sagebrush leaves

1¼ pounds (567 g) dark brown sugar, or 1 pound (454 g) dry malt extract

Zest of 1 orange

Commercial beer yeast or wild yeast starter

Procedure (see Hot Brewing on page 81)

1. Mix the water, sagebrush, and sugar in a large pot. Bring the solution to a boil; let it boil for 30 minutes. Add the orange zest 5 minutes before the end of the boiling time.

2. Remove the pot from the heat and place it in a pan or sink full of cold water. Cool to 70°F (21°C), then add the yeast (wild or commercial). When I'm using a wild yeast starter, I usually use ½ to ¾ cup (120–180 ml) of liquid.

3. Strain the brew into the fermenter. Position the airlock or cover your fermenter with a paper towel or cheesecloth. Let the beer ferment for 10 days. Start counting when the fermentation is active (this may take 2 to 3 days with a wild yeast starter).

4. Siphon into 16-ounce (500 ml) swing-top beer bottles (you'll need seven bottles) and prime each one with ½ teaspoon (2 g) white or brown sugar for carbonation. Close the bottles and store in a place that's not too hot. The beer will be ready to drink in 3 to 4 weeks. If you used dry malt extract, you can do a full fermentation then exact priming to achieve the carbonation you want (see "Civilized Carbonation in Bottles" on page 100).

The Wildcrafting Brewer

YERBA SANTA–MANZANITA BEER

Yerba santa (*Eriodictyon californicum*) is found in Oregon and California. Traditionally, the plant has been used in medicinal teas for respiratory conditions and as a tonic. The plant is also used as a flavoring in the food industry. Most of the flavors are located in the sticky coating around the young green leaves in spring or summer. They're best extracted in a cold infusion or cold fermentation. Boiled as a medicinal tea, the plant is quite bitter, which prompted me to use it as a possible hops substitute. I also love to use the plant in sodas or wines using the cold fermentation method. Manzanita is a local berry that tastes like apple or, in the case of sticky manzanita, like lemon.

1 gallon (3.78 L) water
1¼ pounds (567 g) dark brown sugar
1.2 ounces (34 g) fresh young yerba santa tops
2 ounces (56 g) large sticky manzanita berries
Commercial beer yeast or wild yeast starter

Procedure (see Hot Brewing on page 81)

1. Mix the water, brown sugar, yerba santa, and manzanita berries in a large pot. Bring the solution to a boil; let it boil for 30 minutes.

2. Remove the pot from the heat and place it in a pan or sink full of cold water. Cool to 70°F (21°C), then add the yeast (wild or commercial). When I'm using a wild yeast starter, I usually use ½ to ¾ cup (120–180 ml) of liquid.

3. Strain the brew into the fermenter. Position the airlock or cover your fermenter with a paper towel or cheesecloth. Let the beer ferment for 10 days. Start counting when the fermentation is active (this may take 2 to 3 days with a wild yeast starter).

4. Siphon into 16-ounce (500 ml) swing-top beer bottles (you'll need seven bottles) and prime each bottle with ½ teaspoon (2 g) brown sugar for carbonation. Close the bottles and store in a place that's not too hot. The beer will be ready to drink in 3 to 4 weeks.

WHITE SAGE
(OR CULINARY SAGE) CIDER

White sage can be found in the chaparral and high desert areas of the southwestern United States and northern Mexico. The plant is considered sacred in many native cultures and was used as a food source, for tea, as herbal medicine (it has antibacterial properties), and also in purification rituals. It's much better to plant white sage in your garden than to forage it due to its overuse as smudge sticks in New Age ceremonies. Because of its strong flavors, using white sage is truly an exercise in moderation. In this recipe, the flavor reminds me more of a cider than a beer, hence I call it a cider, despite the fact that it doesn't contain any trace of apples. It's quite delicious.

Be aware that white sage and many other (culinary) sages do contain thujone, so drink in moderation and on special occasions. The recipe uses the same amount as many sage medicinal teas—around one leaf or less per 1 to 2 cups of water.

1 gallon water (3.78 L)

0.10–0.15 ounce (3–4 g) dried white sage or culinary sage leaves

1¼ pounds (567 g) light brown sugar

3 large limes

Yeast (beer yeast or wild yeast)

Procedure

1. Mix the water, sage, and brown sugar; cut and squeeze the limes into the solution. Boil in a large pot for 30 minutes.

2. Remove the pot from the heat and place it in a pan or sink full of cold water. Cool to 70°F (21°C), then add the yeast (wild or commercial). When I'm using a wild yeast starter, I usually use ½ to ¾ cup (120–180 ml) of liquid.

3. Strain the brew into the fermenter. Position the airlock or cover your fermenter with a paper towel or cheesecloth. Let the beer ferment for 10 days. Start counting when the fermentation is active (this may take 2 to 3 days with a wild yeast starter).

4. Siphon into 16-ounce (500 ml) swing-top beer bottles (you'll need seven bottles) and prime each bottle with ½ teaspoon (2 g) brown sugar for carbonation. Close the bottles and store in a place that's not too hot. The beer will be ready to drink in 3 to 4 weeks.

   Based on my experience and depending on the time of the year, the drink can be overcarbonated, so monitor the fermentation and open the bottles carefully.

Brewing as a Representation of Your Own Terroir

Each environment has its unique flora and ecosystem. My location in Los Angeles is unusual in the sense that within an hour's drive I have access to several different types of ecosystem, such as chaparral, desert, mountains, coastal areas, and forests.

From a culinary perspective (which includes fermented drinks), if you know enough about the edible plants and animals living in a specific region, you can create a cuisine representing it. The process may be a bit harder in the desert—or extremely easy in coastal areas, if you take into account the bounty from the sea.

Around five years ago I started researching the possibility of creating wild primitive brews using only foraged ingredients from my local forest and have since experimented quite extensively with whole environments such as my local chaparral, mountains, or desert. As I explained in my book *The New Wildcrafted Cuisine*, the original inspiration came during a forest hike after a couple of rainy days. The whole forest had an incredible "winter" perfume emanating from the countless plants, trees, mushrooms, and leaves decomposing on the ground. It was like smelling the wilderness itself, and I was inspired to re-create the experience from a culinary perspective.

In the beginning I started introducing forest ingredients into my cooking, and I really liked some of the results. But as I explored native and primitive brews, I became fascinated with using some of the same ingredients in my fermentations.

Forests—or other environments, such as the mountains, desert, or the local chaparral—are an endless source of flavors. Sometimes it takes a lot of research to make sure you can use the various ingredients safely, but as you find more and more to play with, you'll learn that this is truly a worthwhile activity.

Do these brews really taste like the forest, the mountains, or other locations?

Yes and no. We're entering a subjective domain. If you asked someone, "What should a forest taste like?," you would get a lot of different answers, but I think that if you smelled or tasted some of my concoctions you would know where they came from. It's much easier with the mountains, where you have very distinctive aromas, such as pine or fir. In my experience a typical forest is a much more complex environment, but you can indeed make beverages representative of your local environment by using flavorful ingredients found within it.

Presently my forest brews are most closely related to Belgian sour beers; very often they end up tasting somewhere between "like a beer" and "like a

cider." They're not something you want to drink in large quantity like regular beers, but based on the feedback I've received, they're quite enjoyable. Some of them are fantastic for cooking.

The recipes and flavors are always changing with the seasons, too. Some plants such as mugwort will have a stronger flavor once the plant has set seeds. In summer dry mugwort leaves will have tons of character if foraged in the shade, though in the sun they are sometimes pretty much tasteless. The last beer I made with seasonal forest ingredients had interesting accents of grapefruit, although none was present in the recipe. There are always surprises, and as I continue tweaking recipes, I'm able to have more good results and fewer failed attempts.

The quest to make fermented beverages that taste like whole environments or specific locations is really a work in progress, and you could probably spend an entire lifetime experimenting with such wild brews. I find it extremely artistic and creative.

This forest in Vermont was used to make a primitive beer composed of such ingredients as white pine, wild sassafras, spruce, yellow birch branches and bark, maple syrup, and wild yeast.

Fermenting a Beverage Representing an Environment

Over the years I've developed a sort of methodology for creating brews based on specific locations. It works for whole environments such as "the Mountains" or smaller ones like "Gloria's Goat Farm in the Mountains."

I basically do three things:

1. Survey the area.
2. Establish the flavor characteristics.
3. Determine the "essence" of a place.

Surveying is pretty easy: Take hikes and observe what is growing and living in that area. You can observe, take notes and lots of photos, then, once you're back home, research what you found through books, online plant identification groups, websites, and so on. There are usually a lot of such plant identification groups or photos of local flora (with names) on social media sites such as Facebook, Pinterest, Instagram, and so on. Once you have the Latin name of a plant, it's much easier to research it. Always make sure it's okay to collect plants at your site, and always check whether a plant is protected or rare; in such cases it's much better to work with local native plant nurseries than to collect from the wild.

The more hikes you take, the better. In the beginning you may notice all the minute details of a place and which specific plants, mushrooms, trees, and herbs grow there. After a while, and once you know the place well, you'll develop a sense of its base flavors. For example, the closest forest to my home is mainly composed of interesting bitter or aromatic plants and trees such as California sagebrush, mugwort, horehound, black sage, yerba santa, and alder and willow trees, but it has interesting fruity components as well, with feral fig trees and Mexican elder (elderberries) nearby.

Thinking about the type of drinks you want to create may add some focus to your quest. For beer, I'm very interested in those bitter and aromatic herbs, plants, or barks. For wines or sodas, I focus more on the fruits, berries, and aromatic plants. You can create several different types of drinks from the same location, each one being a different creative interpretation of that environment.

Once you're surveyed the area and established its main flavor characteristics, here comes the magic.

The next step in grasping the essence of an environment is a bit unusual: Forget everything you've learned—let it go and clear your mind . . .

Be Zen! Stop the constant flow of thoughts in your head, forget the lists of plants you've made, and put the camera aside. Sit down and just be there. Become one with the environment; let it speak to all your senses: sight, hearing, touch, smell, and taste. It's very much like meditation. Not only is the experience extremely therapeutic in our modern world, but letting the environment talk to you is probably the best way to feel and comprehend its intrinsic qualities and sacredness. It's extremely spiritual. Then, and only then, can you start the process of interpretation and creation through drinks.

By letting a place talk to you, you'll pick up the subtle notes, the intricate blends or accents you missed before. That hint of sweetness you smell may tell you to add a bit of those manzanita berries or a touch of wild fennel to your brew.

My friend Gloria has a goat farm in the local mountains, and it's all about pinyon pine with sagebrush accents. In the same location or nearby, you can find white fir, manzanita berries, elderberries, scrub oaks, coffee berries, Mormon tea (*Ephedra californica*, not the plant from Asia used for weight loss), California juniper berries, and yarrow. Within a 10-minute drive, mugwort is also abundant.

Overall, though, the essence is definitely pinyon pine and sagebrush. Pinyon is probably my favorite pine. Munch on a needle and you get very interesting tangerine/lemon/pine accents. In my opinion, the best flavors are actually in the branches, so I usually brew with whole small branches and not just needles.

Locally, mountain colors are blue (from the sky) then green and tan. The greens are quite muted. The tans come from various desert plants, while dirt and rocks can turn a muddy red in some locations.

I experience a lot of emotions in the mountains, including a sense of freedom and space. I seem to be floating above it all. If I had to convey those feelings and colors into a beer, I would make it light, not too alcoholic, and give it amber or reddish colors. The bitterness would come from yarrow, and undertones would come from pinyon pine or white fir. I might add some California sycamore bark to lend it some reddish accents. To add some lemony flavors, I would use some of the bigberry manzanita (*Arctostaphylos glauca*) available on the property; the dry skins of the berries have some definite sour qualities.

Of course, the beer would change with the seasons. In summer I would make it more floral with some elderflowers and add some young California juniper berries for extra zing. In winter, when the pines have less flavor, I might add some local oak bark and older white fir needles, which, for me, convey a sense of maturity and timelessness.

The recipe you create and how you mix the ingredients are part of interpreting the essence of place. It's really something personal, and there is nothing wrong if you add fruits, spices, and other flavorful herbs, barks, or

Local forest: bitter (mugwort, horehound, alder, willow, sagebrush). Wise and complex (oak). Sweetness (figs, elderberries). Decomposing soil, fall leaves, mushrooms. Hints of fennel. Humid, dark, trails, streams. Green flavors (grass, chickweed, miner's lettuce). Solitude, birds, cold (winter).

My local mountains: pines (smell), lemon (pine needles), tangerine (white fir), hints of sweetness (manzanita and other mild-tasting wild berries), bitter (yarrow and mugwort). Space, sky, fresh air, clean water, snow, cold, wildness, bears, mountain lions. Green and blue. Sense of freedom, awe, escape. Fires.

Chaparral: sages and aromatic plants. Bitter (California sagebrush, yerba santa), sweet (currant, prickly pears, elderberries), sour and lemony (sumac berries). Spicy (black mustard). Dry, prickly (cactuses and yuccas). Insects, birds, flowers, coyotes, rattlesnakes. Blue, tan, yellow. Survival, drought.

Desert: aromatics (various sages), pine and lemon flavors (juniper berries). Prickly (cactuses). Hot, dry, spiritual. Death, beauty, rocks, sand, space, freedom, wind. Creosote bush (smell of rain in the desert). Snakes, lizards, coyotes. Blue, brown, yellow. Harsh, salty. Flowers, beauty, renewal (spring).

plants to convey your own flavorful interpretation. When I make a fall forest beer, I like to add a couple of candy cap mushrooms: Their flavors give a special accent that, to me, expresses nicely the flavor profile of my local forest.

If 20 people interpreted a place, they would end up with 20 different drinks, but I think you would still find within each drink some true local flavors. In the case of Gloria's farm, it would be hard to move away from using yarrow and pinyon pine.

CHOOSING A BASE FLAVOR

My personal approach is to start with a base flavor and work from there. This works for beers but also for sodas, meads, cold infusions, teas, and other drinks. It's like making sure I have a solid foundation before I build a house on top of it. The house may look beautiful at the end, but without the foundation it could not be built.

Each location that you visit will have one base flavor, and sometimes more than one. Gardens often have a much wider variety of edible or aromatic ingredients, so you may be able to create several drinks that represent your garden by using different base flavors.

When I drive through local mountains, my base flavors can change every 20 minutes. At 8,000 feet (2,400 m) of elevation, I mostly have white fir, ponderosa pine, coulter pine, and incense cedar. We also have interesting plants and berries such as common mugwort (*Artemisia vulgaris*, not our native one) and coffee berries. Going back down to 5,000 feet (1,525 m), there are fields of yarrow, various oaks, elder trees, manzanita berries, and wild currants.

In a car we're not always aware of all the changes, but it's very obvious if you stop and take a walk.

It's much better to start by defining smaller areas. Trying to make a drink out of a large, generic area such as "the local mountains" would be difficult given the wide biodiversity. My local mountain range is probably over 350,000 acres (142,000 ha) in area, with elevations ranging from 1,600 to over 10,000 feet (500–3,050 m).

But you can easily stop anywhere, pick a specific place, and create a drink out of it.

On Gloria's farm, as I mentioned, pinyon pines and yarrows are my base flavors. Yarrow is the main bittering base; pinyon pine comes after that. If instead of making a beer I wanted to make a soda, I could just take pinyon pine as my main base flavor and lemony manzanita as my second base flavor. Yarrow would be an aromatic that I could use (or not) when "building" my soda.

To make a beer, I take into account all that can be found locally. This already gives me a lot of basic ingredients to play with.

Base Flavors (foundation)

Pinyon pine, white fir (pine flavors, tangerine)
Yarrow (bitter)

Other Possible Ingredients, Flavors, and Accents

Various pines (lemony flavors from needles), or lemons
Unripe manzanita berries during springtime (sour and tart)
Sweet manzanita berries (late summer and fall; these taste like apple)
Bigberry manzanita (lemony/sour)
Oak bark or wood chips (bitter but lots of character; can be roasted)
Mugwort (bitter and aromatic, can be found within 3 miles/5 km of
    the property)
Elderflower (floral)
Mormon tea (pleasant, mild, and slightly bitter, with subtle hints
    of mint)
Elderberries, coffee berries (fruity/sugary)
California juniper berries
Desert sage (*Salvia dorrii*)
And so on . . .

There are no real rules, and you don't have to prove anything to anyone. Let it be about fun, creativity, and flavors. Start with simple recipes and just a few ingredients, if you want, then work your way toward more complex blends. To save money, you can also brew small batches. I usually test new ideas by brewing small quantities, usually ½ gallon (1.89 L).

WINTER IN THE FOREST BEER

This recipe is ever-changing with the seasons, but it's a good example of a winter forest beer. It looks very much as if I just took leaves, twigs, and herbs directly from the forest floor itself, but every ingredient, even fall leaves, was carefully chosen and contributes to the flavor profile. My first attempts at creating such beers were a bit so-so, but they have vastly improved over time. Presently, the end result is somewhere between a beer and a cider—a bit sour, like some wild-yeast-fermented Belgium beers, but delicious. I don't think you could match this recipe with your own local forest, but maybe this will inspire you to experiment with what your wild terroir has to offer.

1 gallon (3.78 L) water
0.2 ounce (6 g) mixed fall leaves
   (cottonwood, alder, and willow)
0.2 ounce (6 g) forest grass—
   regular grass growing in the forest
1 ounce (28 g) manzanita berries
0.1 ounce (3 g) California sagebrush
0.2 ounce (6 g) dried mugwort leaves
0.3 ounce (9 g) turkey tail mushrooms
1¼ pounds (567 g) dark brown sugar
3 large lemons
Commercial beer yeast or wild yeast starter

Procedure

1. Combine all the ingredients except the lemons and yeast. Cut and squeeze the lemons into the solution. Bring to a boil in a large pot for 30 minutes. Remove from the heat and cool to 70°F (21°C), then add the yeast. When I'm using a wild yeast starter, I usually use ½ to ¾ cup (120–180 ml) of liquid.
2. Strain into the fermenter. Place the airlock or cover with a paper towel or cloth and let it ferment for 10 days. Start counting when the fermentation is active (this may take 2 to 3 days with a wild yeast starter).
3. Siphon into 16-ounce (500 ml) swing-top beer bottles (you'll need seven bottles) and prime each one with ½ teaspoon (2 g) white or brown sugar for carbonation. Close the bottles and store in a place that's not too hot. The beer will be ready to drink in 3 to 4 weeks.

Would you like to explore the smell of the forest after the rain, or the sweet fragrance of its rich soil? Maybe add a small amount of candy cap mushroom or bitter decomposing leaves (willow, alder). I like using turkey tail mushrooms, not just for their bitter flavors but also for their medicinal qualities.

LATE-SPRING MOUNTAIN BEER

Spring is a wonderful time to explore the local mountains. Once the winter snow melts, everything seems to emerge from a long sleep, rejuvenated and full of young and fresh flavors. From my perspective, it's all about pine and lemony sour accents. Even the young pine branches often have strong citrus-like qualities. I can't find lemons in these mountains, but unripe manzanita berries are an excellent substitute. It's a very simple recipe, but a good representation of my local flavors. You can probably do something similar in your area with other unripe berries and local pine, fir, or spruce tips.

1½ gallons (5.7 L) water
0.4 ounce (11 g) dry mugwort
0.1 ounce (3 g) yarrow flower heads
1¾ pounds (794 g) brown sugar
2 cups (500 ml) crushed green
manzanita berries
1 small pinyon pine branch
Wild yeast from unripe pinyon pinecones (or
wild yeast starter)

Pinyon pinecones used as yeast starter.

Procedure (see Hot-and-Cold Brewing, page 87)

1. Add all the ingredients to a pot, aside from the pinyon pinecones and branches, and boil for 30 minutes. Cool the solution to 70°F (21°C) and pour it into a fermenting bucket equipped with an airlock. At this point, I add two or three unripe pinecones for yeast and a small pinyon pine branch (cut the needles). Leave the pine ingredients in the buckets until the fermentation is well started, usually 3 to 4 days, but you could leave them longer for more pronounced pine flavors.
2. Once the fermentation is going well, let it ferment for 9 days. Pour it into 16-ounce (500 ml) swing-top beer bottles, and prime each bottle with ½ teaspoon (2 g) sugar. Close the bottles and store in a place that's not too hot. The beer will be ready to drink in 3 to 4 weeks.

SPRING CHAPARRAL BEER

Chaparral is a typical landscape of California and Mexico; I think of it as the land between the desert and the mountains. It's mostly composed of drought-tolerant plants, which vary quite a bit depending on your location (California is a big state). Locally, I find a lot of cactuses, sages, oak trees, yuccas, yerba santa, California sagebrush, Mexican elder, wild currants, and all kinds of interesting greens when we have some good rain. This beer is a good representation of what can be found in May. The concoction is very floral and fruity due to the elderflowers and the pineapple weed flowers (*Matricaria discoidea*), which taste like, well, pineapple. It's a very welcome, refreshing brew during the hot month of June.

1 gallon (3.78 L) water
0.8 ounce (23 g) fresh yerba santa leaves
0.1 ounce (3 g) California sagebrush
1 cup (250 ml) sumac berries (lemonade berries or sugar bush berries)
1¼ pounds (567 g) light brown sugar
30 heads Mexican elderflowers
2 sprigs woolly bluecurls flowers (optional)
2 cups (500 ml) pineapple weed flowers
Yeast (wild or commercial)

Procedure

1. Place the water, yerba santa, sagebrush, sumac berries, and sugar into a large pot. Bring to a boil for 30 minutes and remove from the heat. Add the elderflowers, woolly bluecurls flowers, pineapple weed flowers, and yeast. Cover the pot and set aside for 24 hours.

2. Strain the brew into the fermenter (bottle or bucket) and add the yeast. When I'm using a wild yeast starter, I usually use ½ to ¾ cup (120–180 ml) of liquid. Position the airlock or cover your fermenter with a paper towel or cheesecloth. Let the beer ferment for 10 days. Start counting when the fermentation is active (this may take 2 to 3 days with a wild yeast starter).

3. Siphon into 16-ounce (500 ml) swing-top bottles and prime each bottle with ½ teaspoon (2 g) brown sugar for carbonation. Close the bottles and store in a place that's not too hot. The beer will be ready to drink in around 3 weeks.

You should be able to find what you need to make beer with what nature offers in your area. Another alternative is to grow a (wild) beer garden with all the necessary plants. It's a project I'm currently working on with local native plants and some I can't find locally. These days it's easy, since I can readily purchase seeds online.

If you live in a more temperate climate, you can explore growing your own hops. I have a friend who even manages to grow them locally in Los Angeles.

There are countless fermented experiments and flavors you can play with. As I wrote earlier, if you use brown sugar or molasses you are pretty much guaranteed to end up with a drink that tastes somewhere between a beer and a cider. Bitter ingredients tend to move the beverage toward more beer-like flavors, while more neutral or sweeter ones bring it toward a cider taste. Usually I leave white sugar and honey for wines, sodas, and meads, but as usual there are exceptions. *Tepache*, a refreshing Mexican drink akin to a low-alcohol wine, uses brown sugar.

A lot of the fun is in exploring more exotic ingredients such as barks, mushrooms, and wild berries that you can't always grow in a garden. If you're not a botanist, herbalist, or experienced forager, you may need to do a bit of study, but it really doesn't take that much time to get started. Whenever I travel to a location I'm not familiar with, I always spend some time online researching local edible or aromatic plants. I also look for herbalists or foragers living in the area and contact them. We're really a big family and, more often than not, they're happy to introduce me to the local flora.

I always check a new place for basic bitter plants—yarrow, horehound, Saint-John's-wort, mugwort, ground ivy, dandelion roots—but if you have an experienced plant person with you, they may introduce you to fantastic new ingredients. If someone was looking for bitter but aromatic brewing herbs in my area, for instance, I would definitely recommend my local yerba santa and California sagebrush.

A little tip for choosing plants and other ingredients to make beer is to approach as much as possible the flavors of hops. That's what most people currently associate with beer. Hops are terribly bitter but also highly aromatic, with zesty citric flavors. It's hard to beat all those flavors in one plant, but by making a good blend of various ingredients you may have locally, you can create a similar flavor profile.

For example, horehound is extremely bitter but otherwise quite bland tasting. You can add lemons or other citrusy ingredients (sour grass, sumac berries, tamarind, unripe berries of various kinds, and so on). Locally, I add a bit of aromatic sage or California sagebrush and voilà!—I end up with a decent brew.

So take a few hikes, meet local people and share information, post photos on plant identification groups online, and start the methodology for creating a beverage representing an environment. It doesn't take long to get started.

By the way, as I said earlier, you don't have to break the law for this type of research and experimentation. If you can make beers, wines, or other similar fermented beverages, you can make a lot of friends quickly. By simply showing up with some of my beers and meeting people, I have gained access to perhaps 2,000 acres (800 ha) of private properties representing each of my local bioregions. The property owners let me teach, research, collect, and even re-introduce native plants.

In some areas (such as natural preserves), of course, picking or removing plants may be prohibited. Public lands that are administered by the federal Bureau of Land Management and some national forests usually allow you to pick plants or berries for your own use (which can include research), but always check first.

Vermont Forest Beer

Every location is unique, and a huge part of the fun of travel is exploring and researching the sites you visit. As you gain more experience and knowledge of plants, the process becomes easier. Still, the basics remain the same: You need a sugar source, yeast, and plants or fruits for flavors.

One of my favorite experiences was in Vermont. I was invited to teach a class on the uses of wild edibles at Sterling College and frankly had no idea about the local flora over there. If you think about it, aside from Maine, Vermont is probably the most remote state from Southern California in terms of location as well as climate.

On my way to the college for the first time, I remember looking at nature and thinking, "I'm literally in the weeds"—most of the plants and trees I saw were completely unknown to me. Luckily, I arrived 2 days before classes started, and my first order of business was to connect with a local botanist and a couple of plant-savvy landowners. It was a fascinating couple of days. I spent most of my time in the woods with my new friends learning about the local plants, mushrooms, and trees. In the evening I researched the species we found and their potential culinary uses.

My goal, with my students, was to create a complete wild food feast at the end of the week using foraged products from their own land. As part of the feast, I also wanted some interesting fermented beverages such as sodas and at least a local forest beer.

For my beer, I searched desperately for nice aromatic bitter plants, but in the short time I was there I couldn't manage to locate any of my favorites such as yarrow, mugwort, or horehound. Still, I had a plentiful supply of bitter dandelion roots.

At the end of my 2 days, I ended up with the following foraged ingredients: wild sassafras and dandelion roots, yellow birch bark and twigs, white pine, blue spruce, grass, curly dock, and Japanese knotweed. Not much, but it was enough to make an interesting fermentation in the tradition of root beers, thanks to the wild sassafras roots and yellow birch with their licorice-like flavors.

Being in Vermont, my (wild) sugar source was quite obvious: maple syrup! And I was extremely lucky, because one of the projects they have each year at the college is to collect and make maple syrup using traditional methods. So I had the best-quality syrup to play with.

The recipe was super simple. First, we boiled the yellow birch twigs and bark for 4 to 5 hours, to extract as much flavor as possible. We started with around 2½ gallons (9.5 L) of water; after all that boiling, we ended up with around 1½ gallons (5.6 L). The next step was to add around four decent-sized sassafras roots, a couple of dandelion roots, a handful of Japanese knotweed stalks, and some curly dock leaves (lemony flavors), plus 1 cup (250 ml) of beautiful grass and around 1¼ pounds (560 g) of maple syrup. All those ingredients were then boiled for an additional 40 minutes.

Fermenting Vermont beer.

The solution was cooled, and I added a couple of spruce branches and a bit of white pine. The needles were cut so the flavors could be extracted more easily. I added some wild yeast from a starter made with dandelion flowers. The beer was fermented for 7 days then strained and bottled for 1 day in swing-top bottles for carbonation. It ended up being delicious: very similar to the flavors of a commercial root beer soda, but with hints of pine and a higher amount of alcohol. Some of the beer was also used to cook rabbit, which ended up incredibly tasty.

The Wild Flavors of Belgium

During my last trip to Belgium, I decided to find and gather ingredients to make a true regional beer (gruit). In this case it was easier said than done, as my time was limited (5 days). It was early April, and I didn't have anyone who could help me identify the right plants. In fact, I'd left Belgium so long before that I didn't even know which plants were available locally.

A quick online search for wild ingredients used to make traditional Belgian beers/gruits a long time ago indicated that I could potentially find such plants as wild rosemary, bog myrtle, yarrow, wormwood, ground ivy, dandelion roots, sea wormwood (*Artemisia maritima*), wild hops, and horehound. While I was familiar with a few of these, most weren't ingredients I was used to finding in California.

From experience, I decided to search for plants based on their environment, and each day I took a hike for a few hours in the local forests, near rivers, in areas surrounding lakes, and, just for fun, around the beautiful Belgian town where I was staying. Not only did I have a great time learning about local edible and medicinal plants, but I was extremely happy to find traditional brewing ingredients within a couple of days. I ended up using dandelion roots, which were plentiful in the farm fields surrounding the village; common wormwood (*Artemisia vulgaris*) and yarrow, found on the edge of a local river; and ground ivy from the nearby forest and my brother's garden. One of the main area industries was making sugar from sugar beets, so I decided to use the local brown sugar.

I used ground ivy fresh, as most of its lemony/tangy taste tends to fade when the leaves are dehydrated. The yarrow and wormwood had much less flavor than what I'm used to in Southern California, but it was early spring in Belgium, and I know the best time to collect both ingredients would have been during early fall. Instead of lemons, I used four stems of bitter dock (*Rumex obtusifolius*). Rhubarb stems would work, too (though not the leaves, which are toxic).

WILD BELGIAN BEER

This beer ended up being quite nice and mild. It was not as aromatic as my regular mugwort-lemon beer, but it had a nice bitterness. This was my first time using ground ivy; next time I plan to double the amount and experiment with using the dehydrated herb as well.

Having spent my childhood in the local forests, I felt a very strong connection to this brew. It reminded me that, as a wildcrafter, I was not fermenting ingredients just for flavors. Through the process of wild fermentation, I was really preserving moments in time, places, and memories. I'm sure it's a beer my father would have loved to try.

1¼ pounds (567 g) brown sugar
0.2 ounce (6 g) wormwood or mugwort
0.2 ounce (6 g) yarrow
0.3 ounce (9 g) chopped dried
  dandelion roots
3–4 crushed stems of bitter dock,
  each 3–4 inches (7.5–10 cm) long,
  or 1–2 lemons
1 gallon (3.78 L) water
0.5 ounce (14 g) fresh ground ivy
Yeast (wild or commercial)

Procedure

1. Place the sugar, wormwood, yarrow, dandelion roots, and bitter dock stems into a large pot with the water. Bring to a light boil for 30 minutes. Add the fresh ground ivy after 20 minutes of boiling.

2. Remove the brew from the heat and strain it into the fermenter, then add the yeast. When I'm using a wild yeast starter, I usually use ½ to ¾ cup (120–180 ml) of liquid. Position the airlock or cover your fermenter with a paper towel or cheesecloth. Let the beer ferment for 10 days. Start counting when the fermentation is active (this may take 2 to 3 days with a wild yeast starter).

3. Siphon into 16-ounce (500 ml) swing-top beer bottles (you'll need seven bottles) and prime each one with ½ teaspoon (2 g) white or brown sugar for carbonation. Close the bottles and store in a place that's not too hot. The beer will be ready to drink in 3 to 4 weeks.

Plants Used in My Wild Belgian Beer

Ground ivy (*Glechoma hederacea*) is an herb used to flavor gruits and beers. Not too bitter but quite aromatic. The smell reminds me a bit of some mild hops.

Common wormwood or mugwort (*Artemisia vulgaris*). A bitter and aromatic herb traditionally used instead of hops.

Dandelion roots (*Taraxacum officinale*). I used the roots as a bittering agent.

Yarrow (*Achillea millefolium*), bitter and aromatic, is also a traditional herb used to flavor beers.

Barks, Twigs, Stems, Seeds, Spices, Leaves, Roots, and Insects

Another good exercise in the Zen of local flavors is to take the time to find new uses for what surrounds us in nature. Most of the time we're wandering along looking for the usual green plants, juicy fruits, or tasty berries, and we're often missing a lot of fascinating ingredients. So from time to time, I do an unusual exercise in which I sit down for a specific amount of time with the goal of finding and investigating something new I could use to brew or cook with. I've learned so much by doing this simple exercise. This is how I came up with the idea of investigating local barks, and I soon found out that white oak, yellow birch, willow, and alder barks were used as bittering or flavoring agents in some traditional old beers. But I have barely scratched the surface of possibilities. What about ash, cottonwood, sycamore, and countless other trees?

Mushrooms are fun, too. Many mushrooms growing on trees are medicinal, with woodsy bitter flavors that could be perfect for brewing. You can find much more than the usual turkey tails, chaga, or reishi. Try searching for birch polypore (*Piptoporus betulinus*), artist's conk (*Ganoderma applanatum*), and tinder polypore or hoof fungus (*Fomes fomentarius*), to name a few. Make sure to properly identify any new mushroom, and to thoroughly research possible allergies or other potential negative issues.

Stems are quite interesting as well, so don't neglect them. Whenever I forage a plant, out of respect I try not to let anything to go to waste. Very often I keep the stem if there is a possible use for it, which is often the case with aromatic plants. At home, I have a collection of probably 20 aromatic stems that I use for cooking. One of my favorite dishes is local trout stuffed with sweet white clover, watercress, lemon, and garlic, then placed in a rudimentary mat made of California sagebrush and black sage stems tied together with yucca string or grass. Roasted in the oven, it's super delicious, and my whole house smells like the local chaparral. But I also like to use these stems in some of my wild beers.

Autumn leaves are a good source of bittering ingredients. Locally I use tree leaves such as willow, oak, or alder and bitter catkins in springtime.

You can add all kinds of woodsy and smoky complexities to your brews by putting wood or roasted wood inside your fermenter. If you think about it, that's why many wines and even beers are aged in wood barrels. I often use oak bark from fallen trees, taking it home and cleaning it. My next step is to pasteurize it by placing it in the oven at 220°F (104°C) for 20 minutes. Then I use a kitchen torch to toast it before placing it in my fermenting beer.

What about insects? When I was doing some research on culinary uses of local insects, I came across a couple of ant species that had very complex

Elderflower and ants beer. This was an experimental beer using ants for their lemony flavors.

lemony flavors. Three years ago I did a very successful experiment replacing my usual lemons with foraged ants. Who knows what other flavors you can find out there?

And of course there are bitter or aromatic and flavorful seeds such as wild fennel, coriander, wild carrot (Queen Anne's lace), caraway, hemp, and so on. Some local wild seeds such as the smoothstem blazing star (*Mentzelia laevicaulis*) even taste like peanut butter once roasted or fried.

So take the time to stop, sit down, and think about new potential ingredients. It's an endless but tasty and fun exploration that's worth getting into.

WOODSY MUSHROOM BEER

Winter is our season for mushrooms in Southern California. My two favorites for brewing are turkey tail mushrooms (bitter and with excellent medicinal properties) and candy cap mushrooms, which have maple-like aromas but must be used in moderation. Too much of them is just overwhelming.

1 gallon (3.78 L) springwater—don't use tap water, which may contain chlorine

0.3 ounce (9 g) dried mugwort

2 medium-sized candy cap mushrooms

0.2 ounce (6 g) hops flowers

10–12 turkey tail mushrooms

3 California sagebrush stems (4–6 inches/10–15 cm long)

2 *loomi* (fermented/dried lemons), crushed (see page 168)

1¼ pounds (567 g) brown sugar

Yeast (wild or commercial)

0.7–1 ounce (20–28 g) toasted oak bark (or use your favorite wood chips)

Procedure (see Hot Brewing on page 81)

1. Place all your ingredients (except the yeast and bark) in a large pot. Bring the solution to a light boil for 30 minutes.
2. Remove the pot from the heat and place it (with the lid on) in cold water. Cool to 70°F (21°C), then add the yeast (wild or commercial). When I'm using a wild yeast starter, I usually use ½ to ¾ cup (120–180 ml) of liquid.
3. Strain the brew into the fermenter and add the roasted bark or wood chips. Position the airlock or cover your fermenter with a paper towel or cheesecloth. Let the brew ferment for 10 days. Start counting when the fermentation is active (this may take 2 to 3 days with a wild yeast starter).
4. Siphon into 16-ounce (500 ml) swing-top beer bottles (you'll need seven bottles) and prime each one with ½ teaspoon (2 g) white or brown sugar for carbonation. Close the bottles and store in a place that's not too hot. The beer will be ready to drink in 3 to 4 weeks.

Black Lemons (Loomi)

As you read this book, you must be thinking that I really *love* lemons. In truth you don't have to use lemons in the majority of the recipes, but I find that adding lemony flavors when using wild bitter plants such as yarrow and horehound instead of hops creates a nice balance.

Hops have a citrus quality that can be lacking in other wild bitter plants. Adding sour/lemony ingredients helps make a wild beer taste a bit more like a regular beer . . . sort of. Not that I'm trying to make regular beer; I think each of my primitive brews is beautiful in its unique way.

You're not stuck with lemons, either. A lot of ingredients have citrus qualities and are used in this book: tamarind, unripe berries, sumac berries, rhubarb stalks, wild sorrel, and so on.

One of my favorite replacements for regular lemons—one that contains unique and complex citrusy tastes and aromas—is the black lemon (aka loomi). They're actually not lemons but whole dehydrated limes. The flavors are very intense, and I think a bit of fermentation goes on inside them, with all their sugar, given the time it takes to dehydrate them (usually a couple of weeks, though the process is faster in the heat of summer in California). In the Middle East they're also used to make a sort of soda.

It's almost impossible to find an exact recipe for making loomi. Most of the recipes I have unearthed are vague about the method. Basically, you boil limes in salt water and dry them in the sun, but I've never learned how much salt to use, how long the boiling time should be, and so on. In desperation, I finally asked my Armenian neighbor if she knew how to make them and her answer was: "Yes! Just leave them in the sun. No boiling necessary. I've done it many times!"

And then I realized why the recipes are vague . . . because it seems everyone has a slightly different way to accomplish this very simple process.

So should you boil? Yes, you should. By doing so, you remove surface yeast and bacteria that could potentially mold or spoil your limes. I boil my limes for 4 to 5 minutes in salted water (1 tablespoon salt per cup of water/17 g salt for 250 ml of water) then place them in the hot sun until they're dried.

Any brew that features mushrooms or evergreens (pine, spruce, or fir) goes extremely well with loomi. I think they also add some forest qualities to a wild beer.

CHAPTER 7

Primitive and Country Wines and Meads

I f you purchased this book in the hope of creating a fine Cabernet Sauvignon, Merlot, or Bordeaux, I'm afraid you've got the wrong book. But if you're looking for adventurous fermented concoctions and interesting new or unusual flavors, keep on reading. Just don't try to compare these wines to the usual commercial products. They are completely distinctive and unique.

Over the years I've made all kinds of primitive drinks and fermented concoctions using garden-grown and foraged ingredients—including berries, wild fruits, aromatic herbs, tree leaves, barks, insects, and natural sugar sources such as tree sap, insect honeydew, fruit molasses, honey, natural yeasts, and so on. And I've come to the conclusion that many native, ancient, and simply made foraged drinks have flavors that are, well... primitive. If you don't believe me, try some pulque (agave "beer" that's drunk as is, but also distilled into tequila) in Mexico.

By the way, when I say "primitive" I don't mean unrefined, rude, or unsophisticated. I mean it in the sense of primal, unaffected, simple, and natural. For me it's a not a bad thing—I think we're rediscovering long-lost flavors, the same ones our ancestors enjoyed a long time ago. That's what's so beautiful about these drinks.

If you're making a primitive or country wine using what your garden or nature has to offer, it's not going to taste like something you can buy at the store. Think of it as traveling back in time. To a large degree we have forgotten the sacred origins of fermentation and primitive brews; their spiritual effects on the psyche or the body were historically valued far more than their flavors or the chance to get drunk. The ancients also knew something that is now largely forgotten: Plants can communicate with us

Fermenting elderberry wine.

through primitive brews on a spiritual or medicinal level—and yes, through taste as well.

The flavors aren't for everyone. Some of the chefs I've worked with didn't know how to deal with them, and I don't blame them. My foraged grape wine brewed with wild yeasts does not compare to a fine Bordeaux or Merlot—though to be fair I'm not trying to. My brew is what it is. After centuries (or more) of winemaking experimentation—altering the raw ingredients such as grape varieties, yeasts, and even sugar—we humans have managed to produce extremely sophisticated and delicate alcoholic beverages. We have also turned the Near Eastern wildcat into a cute purring creature and wolves into tame and friendly dogs. Where I live, if you create with raw nature, even in the present day you will still deal with rattlesnakes, bobcats, mountain lions, black bears, and coyotes. I like to think of my brews as a reflection of a raw, untamed terroir.

On the plus side, though, there is today a definite trend toward rediscovering those long-lost raw flavors. In the beer world, natural yeasts and wild ingredients are making a very strong comeback, and sour beers are becoming popular again.

In the hands of talented and creative chefs or cocktail artists, the primal qualities these brews offer can really shine. Here in Los Angeles, genius mixologist Matthew Biancaniello can pretty much turn any of my foraged and primitive fermented concoctions into drinkable works of art. Chef Josiah Citrin at Melisse restaurant can transform my white sage cider into a delicate sorbet. And ever since my first book, *The New Wildcrafted Cuisine*, was published, quite a few restaurants around the country have begun experimenting with brewing using local plants, cooking with the resulting beverages, and making interesting vinegars.

From a modern culinary perspective, natural fermented drinks, with their raw and undomesticated character, offer a potential as vast as that of the wilderness around us. There is much to explore.

What Is a Country Wine?

The definition of *country wine* is quite wide ranging. You usually find two definitions, as follows:

> **COUNTRY WINE OR VIN DE PAYS:**
> (1) an official French wine-quality classification that falls above table wine but below quality wine.
> (2) wine made from something other than grapes, such as fruit, flowers, or herbs.

I mostly use the second definition: If you use something other than grapes to make wine, you're making a country wine. The term may have originated from the fact that in pretty much any culture or location, it's traditional for people living outside cities to make their own wines using what they have in their gardens and orchards, or from foraging the local environment. Winemaking is simply part of securing the harvest, just like other methods of food preservation.

Here in Los Angeles my local Middle Eastern market stocks all kinds of interesting wines made with black currants, cherries, plums, apricots, and so on. Of course grapes are very juicy and have the right amount of sugar for wine; other fruits, berries, and plants contain less water and may not have enough sugar to create the typical alcoholic content of a wine. Thus, many recipes for country wines require the addition of sugar and water.

Most modern recipes for making country wines call for Campden tablets (potassium metabisulfite) to sterilize the wine, as well as commercial yeast strains, yeast nutrients, and so on. I prefer to rely on the old methods and their simple ingredients: fruits or herbs, wild yeast, organic sugar, water, and nothing else. I think that if you really want to taste your terroir, it's better to use local yeast. But these are all personal choices.

If you're trying a local ingredient that's not in this book, this rule of thumb could help: Typically, country wine recipes call for around 60 percent fruit or berries and 40 percent white sugar. For example, it you're making a black currant wine, for 3 pounds (1.5 kg) of berries you would use around 2 pounds (1 kg) of white sugar. For a sweeter wine, you would use 2½ to 3 pounds (1.25–1.5 kg) of sugar.

Most recipes ask you to boil the juice or the liquid to be fermented, cool the solution, and add the yeast. Because I use wild yeast and boiling would kill it, I often simply ferment the raw juice. Some people might say that fermenting raw juice makes the wine more prone to spoilage or turning into vinegar, but honestly I have never had problems with this method. You also have the option of boiling the solution, cooling it down, and then adding some wild yeast starter.

Most country wines require 6 months to a year of aging to develop optimal flavors.

What Is a Primitive Wine?

Go back 10,000 years, and wines or alcoholic beverages were made with what was found locally. Foraged wild fruits or berries were crushed and mixed with fresh springwater; then honey or another source of sugar such as fruit molasses or possibly insect honeydew was added to increase alcohol content. Herbs could also be used for flavors, for their psychotropic effects, or for their medicinal properties.

Because it wasn't boiled, the mixture of ingredients created a spontaneous fermentation due to the presence of wild yeasts on the fruits or berries, or simply in the air. Such a beverage would be drunk within a few days, before it had the opportunity to turn into vinegar or become too sour. To this day the Tohono O'odham people in the Sonoran Desert continue a tradition of fermenting this type of beverage with saguaro cactus pears. In my opinion, primitive wines are the most fun simply because they're extremely easy to make and perfect for experimenting with local ingredients.

Juicing wild currant berries to make wine.

BASIC COUNTRY WINE—
FRESH BERRIES OR FRUITS

If you ever find yourself in the middle of nowhere, with no Internet access to research specific recipes, and you have tons of local fruits or berries you want to use, here is the basic procedure for turning them into country wine. With experience, you can tweak this recipe a bit, but as is it will provide you an enjoyable alcoholic beverage.

If you use wild yeast (which usually won't tolerate a high level of alcohol), I suggest you use the lesser amount of sugar. Very often I also taste the wine during the fermentation process and add more sugar if necessary.

3 pounds (1.5 kg) berries or fruits (black currants, blueberries, raspberries, blackberries, etc.)
1 gallon (3.78 L) water
2–3 pounds (1–1.5 kg) granulated white sugar (using more sugar will make the wine sweeter; realize you can always add sugar to a fermented wine)
½–1 teaspoon (2.5–5 g) citric acid, or juice of 2–3 lemons (adds fruity flavors and also helps with preservation)
Wine yeast or wild yeast

Procedure

1. Thoroughly clean your hands and any instruments or materials you will use. Maintain this high level of hygiene throughout the winemaking process.
2. Strip and rinse the berries, and remove the pits from the fruits. Using a large bowl and your clean hands (or a potato masher), crush or mash the berries/fruits and place them into your fermenting container (a large jar, bucket, or the like).
3. Boil the water and pour it over the berries or fruits. Add the sugar and citric acid; if you want, you can also add some aromatic herbs at this stage. Stir and make sure the sugar is dissolved.

4. Cover and leave the mixture for a day or two, stirring at least three times daily with a clean spoon. Some people add the yeast at this stage (after the liquid has cooled to room temperature).
5. Strain the liquid into a fermenting bucket or bottle with an airlock and add the yeast. When I'm using a wild yeast starter, I usually use ½ to ¾ cup (120–180 ml) of liquid per gallon. Place your container in a cool, dark place until the fermentation is complete, usually 3 to 4 weeks.
6. Some people rack the wine after a couple of weeks. Racking is the careful transfer of wine from one fermenting container to another, which helps clarify it and remove sediments. I usually don't rack my wine; I simply bottle it carefully after full fermentation, making sure the sediments at the bottom are not disturbed.
7. When the fermentation is complete, bottle the wine. I've been known to leave my wine for 3 to 4 months in the original fermenting container before bottling it. Make sure the liquid in the airlock does not evaporate, and refill it with water if it does.

The wine can be drunk after 6 months, but it's much better after a year (or more).

RAW WINE METHOD

These days, I tend to ferment mostly raw wines. It's the same recipe as the basic country wine, but I don't boil the water.

Procedure

1. With clean hands, remove the berries from the stems, crush them, and in a large container combine them with the water, sugar, and citric acid. Cover the container with a clean towel and leave alone for 5 to 6 hours. This will allow the yeast that was present on the skin to interact with the sugary liquid.

2. Strain the juice through a cheesecloth or sieve into your fermenter (large bottle). If the berries you used had a nice bloom/wild yeast on them (blueberries, wild grapes, et cetera), there is no need to add yeast; It's already inside the juice. If you're unsure, add some wild yeast starter. If the fermentation doesn't start within 2 to 3 days, you can add some starter at that point as well.

   Very often I'll also add (cleaned) aromatic herbs such as mugwort, yarrow, et cetera, in the fermenter at that stage for added flavors.

3. After 2 to 3 days of active fermentation, I usually remove my aromatic herbs, strain the contents (you usually have floating particles) into a new bottle, and let it ferment until the fermentation is complete, then bottle it for 6 months or so.

   Using the raw method, your wine can be more prone to spoiling (often turning into yummy vinegar)—but honestly, it has never happened to me. Good hygiene and thoroughly cleaning your utensils and containers are key.

MAKING COUNTRY WINE
WITH DRIED BERRIES OR FRUITS

Country wine can be made from dehydrated ingredients in a process that's very similar to the recipe for fresh ingredients. The difference is that you use much less of the dehydrated fruit (around 25 to 30 percent by weight), mostly owing to its absence of water.

1 gallon (3.78 L) water

2–3 pounds (1–1.5 kg) granulated white sugar (more sugar will make the wine sweeter; be aware that you can always add sugar to a fermented wine)

Around 1 pound (454 g) dried berries or fruits (currants, elderberries, blueberries, etc.)

½–1 teaspoon (2.5–5 g) citric acid, or juice of 2–3 lemons (this adds fruity flavors but also helps with preservation)

Wine yeast or wild yeast

Procedure

1. Thoroughly clean your hands and any instruments or materials you will use. Maintain this high level of hygiene throughout the winemaking process.

2. In a large pot, bring the water to a boil and dissolve the sugar in it. Remove from the heat, and add your dried berries and citric acid. You can also add dry aromatic or flavorful herbs at this stage.

3. Cover the pot and leave it for a couple of days. (I usually leave it for 36 hours.) Stir at least three times daily with a clean spoon. Some people add the yeast at this stage (after the liquid has cooled to room temperature).

4. Strain the liquid into a fermenting bucket or bottle fitted with an airlock and add the yeast (if you didn't do so earlier). When I'm using a wild yeast starter, I usually use ½ to ¾ cup (120–180 ml) of liquid. Leave until the fermentation is complete, usually 3 to 4 weeks.

5. Some people rack the wine after a couple of weeks. Racking is the careful transfer of wine from one fermenting container to another, which helps clarify it and remove sediments.

6. When the fermentation is complete, bottle the wine (by using a siphon or, if you made just 1 gallon, by pouring it carefully into the final bottles). I've been known to leave my wine for 3 to 4 months in the original fermenting container before bottling. Make sure the liquid in the airlock does not evaporate; if it does, refill it with water.

    The wine can be drunk after 6 months, but it's much better after a year (or more).

OTHER COUNTRY WINE RECIPES

Sweet Currant Wine—Fresh Berries

3 pounds (1.5 kg) black currants or other currant berries

1 gallon (3.78 L) water

3 pounds (1.5 kg) granulated white sugar

½–1 teaspoon (2.5–5 g) citric acid, or juice of 2–3 lemons

Wine yeast (use only 2–2½ pounds/1–1.25 kg sugar if using wild yeast)

Blueberry Wine

2 pounds (1 kg) blueberries

1 pound (454 g) raisins

1 gallon (3.78 L) water

3 pounds (1.5 kg) granulated white sugar

½–1 teaspoon (2.5–5 g) citric acid, or juice of 2–3 lemons

Wine yeast (use only 2–2½ pounds/1–1.25 kg sugar if using wild yeast)

Blackberry Wine

2–3 pounds (1–1.5 kg) fresh blackberries

1 gallon (3.78 L) water

2–2½ pounds (1–1.25 kg) granulated white sugar

½–1 teaspoon (2.5–5 g) citric acid, or juice of 2–3 lemons

Wine yeast (use only 2–2½ pounds/1–1.25 kg sugar if using wild yeast)

Sweet Blueberry Wine

3–4 pounds (1.5–2 kg) fresh blueberries

1 gallon (3.78 L) water

3 pounds (1.5 kg) granulated white sugar

½–1 teaspoon (2.5–5 g) citric acid, or juice of 2–3 lemons.

Wine yeast (use only 2–2½ pounds/1–1.25 kg sugar if using wild yeast)

Procedure

1. Thoroughly clean your hands and any instruments or materials you will use. Maintain this high level of hygiene throughout the winemaking process.

2. Strip the berries from their stems and rinse. Using a large bowl and your clean hands (or a potato masher), crush or mash the berries and place them into your fermenting container (a large jar, bucket, or the like).

3. Boil the water and pour it over the berries. Add the sugar and citric acid; if you like, you can also add some aromatic herbs at this stage. Stir and make sure the sugar is dissolved.

4. Cover the pot and leave for a couple of days. Stir at least three times daily with a clean spoon. Some people add the yeast at this stage (after the liquid has cooled to room temperature).

5. Strain the liquid into a fermenting bucket or bottle fitted with an airlock and add the yeast (if you didn't do so earlier). When I'm using a wild yeast starter, I usually use ½ to ¾ cup (120–180 ml) of liquid. Leave until the fermentation is complete, usually 3 to 4 weeks.

6. Some people rack the wine after a couple of weeks. Racking is the careful transfer of wine from one fermenting container to another, which helps clarify it and remove sediments. I usually don't rack my wine; I simply bottle it carefully, making sure the sediments at the bottom are not disturbed.

7. When the fermentation is complete, bottle the wine (by using a siphon or, if you made just 1 gallon, by pouring it carefully into the final bottles). I've been known to leave my wine for 3 to 4 months in the original fermenting container before bottling it, but only if I used a vessel fitted with an airlock, which protects the wine much more than a container with a towel on top. Make sure the liquid in the airlock does not evaporate; refill the airlock with water if necessary.

The wine can be drunk after 6 months, but it's much better after a year (or more).

CHAPARRAL COUNTRY WINE

Instead of making wines around specific berries, I like to create blends that represent whole environments. My Chaparral Country Wine uses wild currant and manzanita berries, sometimes Mexican elderberries or coffee berries (*Frangula californica*), and a tad of mugwort. A mountain country wine recipe, on the other hand, would include ingredients such as black elderberries, pinyon pine branches, a bit of yarrow, manzanita berries, and California juniper berries.

1 gallon (3.78 L) water

2–3 pounds (1–1.5 kg) granulated white sugar (more sugar will make the wine sweeter, but remember that you can always add sugar to a fermented wine later on; when using wild yeast, I only use 2 pounds/1 kg sugar)

5 ounces (140 g) dehydrated black currants

2 ounces (56 g) manzanita berries

1.5 ounces (35 g) dehydrated coffee berries

3 ounces (84 g) dehydrated red currants

0.1 ounce (3 g) dry mugwort

Wine yeast or wild yeast

½–1 teaspoon (2.5–5 g) citric acid, or juice of 2–3 lemons

Procedure

Use the preparation method detailed in "Making Country Wine with Dried Berries or Fruits" (page 177).

Mixed Wild Berry Recipes

I've found that after a year, many of the plants and dried berries in my pantry start to lose some of their delicate aromatics or taste. Thus it's a good idea to renew your pantry every year for optimal flavors. This renewal is also a great time to experiment with unusual country wine recipes using the older dehydrated ingredients.

Gather all your ingredients: dried berries, herbs, and lemons. Meanwhile, bring 1 gallon (3.78 L) of water to a boil and dissolve the sugar in it.

Creating Your Own Country Wines

Country wines are super easy to make. It doesn't matter where you live; you can always find local ingredients, fruits, or berries to play with. Some people even make country wines with vegetables such as carrots or rhubarb.

So have fun and be creative! Maybe you can start with simple recipes then, as you gain experience, move into experimenting with more complex blends of local berries, barks, roots, and flavorful herbs that truly represent your local terroir.

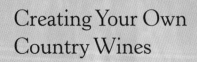

FLOWER WINE
(ELDERFLOWER CHAMPAGNE)

This is an old traditional European recipe for making wine with elderflower. In Southern California we have Mexican elders (*Sambucus mexicana*) at low altitudes and the regular elder (*S. nigra*) in the mountains. One of the peculiarities of the Mexican elder is the fact that the flowers can be smaller, usually half to a quarter the size of the regular elder, which changes the recipe a bit.

I don't know why the wine is called a champagne—perhaps it's due to the color and the fact that it's bubbly. The old recipes make no mention of adding yeast, because it's present on the flowers. I've had moderate success (probably around 70 percent) with spontaneous fermentation from the flowers, so these days I usually add some champagne or wine yeast if I don't see any signs of fermentation after a couple of days.

30 large Mexican elderflower heads,
    or 20 regular elderflower heads
1 gallon (3.78 L) water
3 cups (500–600 g) white sugar
3–4 lemons, zested and sliced
2 tablespoons (30 ml) vinegar (I use apple
    cider vinegar)
Champagne or wine yeast (optional—
    flowers should have wild yeast)

Procedure

1. Pick the elderflowers when they're fresh and full of pollen. Fresh Mexican elderflowers look a bit greenish, while the older flowers are whiter. You'll discover very quickly that elderflowers are loaded with little bugs. My solution to get rid of (most of) them is to place the flowers in a bowl outdoors for about an hour; the little bugs will vacate. You can't really remove them all at this point, but as you strain your solution later on, it will take care of the remaining ones.

2. Place the water in a container, add the sugar, and stir with a clean spoon to make sure it's dissolved.

3. Add the lemon zest and lemon slices, the elderflowers (remove as much of the stems as you can without going crazy about it), and the vinegar to the container and stir briefly with a clean spoon. Some people add commercial yeast at this stage.

4. Close the container, but not so tight that fermentation gases can't escape. You can also place a clean towel on top. Let the mixture stand for anywhere from 24 to 48 hours. If you didn't use yeast, you should see some bubbles after 48 hours, indicating that the fermentation from wild yeast is active. If

this doesn't occur, then add some yeast at this stage. Using a clean spoon, make sure that you stir the liquid for a few seconds three or four times a day during this process.

5. Strain the liquid (after 48 hours if additional yeast was necessary) into your fermenting vessel (bottle or bucket). Let the fermentation go for another 4 days. Using a layered cheesecloth when straining the liquid removes any remaining little bugs.

6. Your final step is to bottle your champagne in recycled soda bottles or swing-top glass bottles. Let it ferment for a week before enjoying. I like to check the pressure from time to time by unscrewing the bottle slightly to make sure it's not excessive.

COUNTRY ELDERBERRY WINE
(COOKED AND RAW)

This is an old French recipe I discovered. It uses a bit of an unusual method, but it's also much simpler, and honestly I like the flavor better. The standard winemaking technique involves boiling and simmering juice, which unfortunately also kills the wild yeast; thus winemakers typically add commercial yeast (wine or champagne). This recipe, however, forgoes the yeast.

California is home to Mexican elder trees (*Sambucus mexicana*), which provide either white or black berries (they look similar to white or black grapes in terms of color). These berries are also more sugary than other elderberries, so if you live in a different state or country and are working with the usual black-berried elder (*S. nigra*), you may want to add a bit more sugar than this recipe calls for. Or you can also just follow the amounts given here, then sample the brew when fermentation is complete and sweeten to match your taste.

Note that you need to pick the berries when they are *fully* ripe for this recipe. Unripe berries may contain cyanidin glycoside, which isn't good for you; drinking raw elderberry juice is also not advised. Fermentation with raw juice is documented in the books *Preserving Food Without Freezing or Canning* by the Gardeners and Farmers of Centre Terre Vivante and *Wild Fermentation* by Sandor Ellix Katz; it's the only method I've used for the last three years. The Nordic Food Lab has also done some interesting work related to the effect of fermentation on cyanogenic glycosides in elderberries, which is worth reading on their website (nordicfoodlab.com).

I've used this (raw) recipe with *Sambucus mexicana* and *S. nigra* (commonly found in Europe and North America). But note that the recipe may not be appropriate for other types of elder trees and berries.

If you're unsure about ripeness or other issues (cyanogenic glycosides, types of elderberries, or the like), simply use the same basic recipe but boil the juice and add commercial wine yeast or wild yeast starter once the solution has cooled down.

Crushing elderberries to extract the juice.

Old French Recipe

55 pounds (25 kg) elderberries
5.5 pounds (2.5 kg) honey
5.5 pounds (2.5 kg) white sugar
5 teaspoons (25 g) salt

Crush all the berries and place them into a barrel with the other ingredients. Let everything rest for 1 month, then strain the wine into bottles.

My Modern Interpretation

4 pounds (2 kg) elderberries
   (*Sambucus nigra* or *S. mexicana*)
2 pounds (1 kg) granulated white sugar
1 gallon (3.78 L) springwater or
   distilled water
1 teaspoon (5 g) citric acid, or
   juice of 3 lemons
Elderberry wild yeast
   (already on the berries)

Procedure

1. With clean hands, remove the berries from the stems, crush them, and strain the juice through a cheesecloth or sieve.
2. In a clean pot, combine your juice with the sugar, water, and citric acid. There's no need to add yeast; it's already inside the juice.
3. Place the mixture in a large (clean) bottle fitted with an airlock and let it ferment for 6 months, then bottle. Wait another 6 months before drinking if you have the patience (which I don't).

This wine should end up a tad sweet. Wild yeast from elderberries can be quite interesting and variable. I've made batches that had a higher percentage of alcohol and were drier. Just taste the brew when it's ready and add sugar if necessary.

Because it uses raw juice and wild yeast, this wine is probably more prone to spoiling (it turns to a yummy, useful vinegar) than wines made with the boiling method. But so far at least, this has never happened to me. Just make sure to clean thoroughly all the equipment you use in the process.

Elderberry Slime, aka Green Goo or Gunk

In the beginning of the fermentation, you should see some sort of green slime or goo on top of your fermenting liquid. Don't worry too much about it—it's pretty normal with elderberry fermentation. It's easy to clean with hot water later on. If you want, you also have the option to transfer (by racking) your wine to another container after a couple of months.

Boiled Wine Variation

Follow the same procedure and on step 2 bring the liquid to a boil then simmer for 10 minutes. Cool the solution by placing your pot in cold water (change the water a couple of times if necessary), then add the yeast (wine yeast or yeast starter). When I'm using a wild yeast starter, I usually use a bit more than ½ cup (120 ml) of liquid.

Lazy Fruit or Berry Wines

When it comes to fermentation methods, you can't get any simpler than this one: Place your fruits or berries into a pot, maybe crush or cut them a bit, add some herbs for flavors, and voilà!—the wild yeast already present does the work. The more I do this type of fermentation, the more I tend to believe that this was how the original primitive wines were created.

The end result is actually similar to a type of fermented drink called fruit kvass, which we'll explore further in chapter 8. I use fewer ingredients, however—usually just one type of watery and sugary fruit, such as plum, prickly pear, or pomegranate, or specific berries like grapes, blueberries, or currants. Then I add to that some flavorful herbs or spices.

Fruit kvass are a bit different, often mainly composed of more fibrous and fleshy fruits like apples, pears, apricots, kiwis, strawberries, and so on. But there are no real rules, either. My "lazy" wine usually has more alcohol than a regular fruit kvass, but it's (usually) not fermented fully like a regular wine. If I had a scale for fermented fruit or berry beverages, it would look like this:

Light fermentation for sodas: 1 percent alcohol or less
Fruit kvass: 1 to 3 percent alcohol
Lazy wine: 3 to 5 percent alcohol
Country wine: 5 percent alcohol and up

I call it this sort of fermented beverage a "lazy wine" because it's so easy to make. About the only effort required is shaking a container a few times daily and maybe adding some sugar during the process. The idea is that you taste your fermenting concoction as you go along, and you can stop anytime. Some people may like a young wine with only a bit of fermentation, while others may be more interested by a more alcoholic drink. I've fermented some wines for just a week and others for a month. In some cases, I might even decide at the end to squeeze and extract the fermented juice, then age it in a separate container.

Here is a good example of the method (such as it is) that I use with grapes. The same method can be used with plums and various other fruits. I chose grapes because they're readily available and a lot of people will try

Organic grapes, yarrow flowers, and roasted alder bark (bitter) from a recent forest fire. Start with the jar 75 percent full. In the photos the berries are already starting to expand and pushing upward. It's wise to set a bowl or tray under the container in case some of the liquid spills over.

Active wild yeast action after 6 days. Stir the ingredients at least three times (or more) daily until you're happy with the flavors. You can also add sugar and other ingredients during the fermentation; there are no hard-and-fast rules in making lazy wines!

them—but as you'll see, commercial fruits from the market can actually be more finicky due to potential yeast issues. I much prefer wild fruits and berries for making lazy wines. (See also "The Zen of Fermenting" on page 104.)

Method

Collect fruits/berries, or in this case grapes (from your garden, foraged from the wild, or from your local farmer's market). Clean them briefly to remove any dirt but not the wild yeast located on the skin. For this wine, I slice the grapes in half so the sugary juice can leach into the water. You can also wash the berries thoroughly and use commercial wine or champagne yeast if you want.

Because I worked with a 1-gallon (3.78 L) container, I used 12 cups (around 2.8 L) of sliced grapes and just enough water to keep the grapes under the surface. It took around 6 cups (same volume as 1.4 liters) of spring-water. You *don't* want to fill up the jar to the top; I usually fill it to 75 percent of its volume. Why? Because the brew generates a large amount of bubbles. This means that your grapes or berries have a tendency to float on top of the liquid and be pushed up (see the photo on page 190). That's not a huge problem if you shake the mixture several times a day. Still, it's better if you don't fill the jar completely to start with. Aside from just the grapes and water, I also added 1 cup (200 g) of organic white sugar and 1 tablespoon (15 ml) of raw honey (optional, but it's an additional source of wild yeast). If your grapes are very sweet, you may not need to add sugar at all.

For added flavors, I included a bit of yarrow (three to four flower heads) and two pieces of roasted alder bark from a recent forest fire.

You might prefer to include aromatic plants, spices, or herbs such as fennel, lavender, some mild sages, peppercorns, cloves, spruce tips, or coriander seeds. If you like bitter accents, you could experiment a bit with hops, mugwort, wormwood, California sagebrush, or other bitter aromatic herbs. I've also used toasted oak chips or many other types of aromatic woods, such as mesquite, hickory, and more.

Close the container, but make sure the fermentation gases can escape. In my case, I left the top of the jar somewhat loose so it could act as a sort of airlock. If the fermentation is quite active, you may need to open the lid a bit and place a clean towel on top for a few days. Of course, you can also simply use a large jar with clean paper towel placed on top and held in place with a rubber band, or—even better—a container fitted with an airlock. It's all good.

You'll need to stir the contents with a clean spoon *at least* three times a day. In the beginning, if a fermentation is super active, I may stir it up to ten times a day to make sure the fermentation gases pushing up the berries don't create a mess. It's a good idea to place a bowl or plate under the jar. Eventually the fermentation will calm down.

Given the amount of wild yeast present on the grape skins, the fermentation usually starts within a day or two at room temperature (70–80°F/21–27°C). You'll see some bubbles showing up on top; within 3 days your fermentation should be active. This is when I start tasting the fermenting liquid every 3 to 4 days using a clean spoon and evaluating whether I need to add sugar, stop the fermentation and store it in the fridge, add some herbs or spices, or try a new creative idea. The idea is to work with the fermenting liquid as you go along and stop when you like it. You can also decide to do a full fermentation if you're more interested in a higher alcohol content.

My own simple rule is to stop when I actually like the flavor, which can take anywhere from 1 to 3 weeks. I then strain the wine and either drink it within a day or two or store it in the refrigerator, which slows down the fermentation. Most of these wines are not meant to be aged, but you can do so if you want.

Every lazy wine has a different fermentation based on the ingredients you use and, of course, the yeast. Commercial grapes are definitely not my favorite for lazy wines, but they do offer a fine learning experience. I made a couple of grape wines that ended up excellent after a bit more than a week of fermentation, still nice and sweet without excessive alcohol.

My latest batch, made with organic seedless grapes (seeds can add a bitter taste) that I purchased at the local farmer's market, behaved entirely differently: It just kept fermenting and fermenting. Despite my years of experience with wild fermentation, I was not prepared for this brew. It went way over the usual "around 5 to 7 percent" alcohol; based on taste, I'd say it reached well above 10 percent ABV at the end. I suspect that it could have been a case of feral yeast—a strain of commercial cultured yeast that has taken over a specific area or winery and replaced the original wild yeast.

These specific grapes were so sweet that I didn't expect to add any sugar at all. After a week, though, the wine started to get sour due to the active fermentation and lack of sugar. I ended up having to add 2 cups (400 g) of sugar (for 1 gallon/3.78 L) within 2 weeks. And still it would not stop fermenting and getting more alcoholic. Finally I strained the fermenting concoction using a mesh bag and poured the liquid into a bottle fitted with an airlock for a full fermentation. I aged it for a couple of months. It ended up not tasting wonderful, but it was highly alcoholic. I used it mostly for cooking and made some decent vinegar out of it. If it is legal to distill alcohol in your country (it isn't in the United States) and you have the equipment, such a brew would be perfect for making a grappa (clear grape-based brandy).

LAZY PRICKLY PEAR WINE

This recipe uses the same technique as the lazy wine made with grapes, though it's even simpler and I have never experienced a feral yeast problem. It's one of my favorite methods to make primitive wine with our local prickly pear cactus fruits, which are hard to juice because they contain a lot of seeds. You could add some aromatic wood, herbs, or plants if you want to, but it's really good without any additions. In California you can also purchase large prickly pears at local Hispanic stores (nopal cactus), but I like to forage the smaller ones from our local species (*Opuntia littoralis*).

There are numerous methods for harvesting prickly pears. I always use my hands and a small broom, usually one I make with twigs found on location. I don't use kitchen tongs, don't burn the needles or anything else. You just need to vigorously slap or brush the fruit up, down, and sideways for 10 to 20 seconds; this will remove 100 percent of the *glochids* (tiny needles). It takes a bit of practice, but it works: I don't get any needles in my skin or fingers. Don't do it with the wind blowing toward you, though, or the flying needles will get into your clothes.

When you've removed the glochids, use a hand twist to remove the pear. Do one more quick inspection of your work and place the pear in the bag. Back at home, I don't wash pears I'll be using for fermentation; the yeast is present on the skin, and a thorough cleaning would remove it. If necessary, a very quick rinse with cold water is okay.

10–12 cactus pears

½ cup (100 g) white sugar, or ½ cup (120 ml) honey, to start with

Enough springwater to fill 90 percent of the jar

Procedure

1. Using a knife, stab each cactus fruit three or four times, and place them all in a clean 1-quart (1 L) jar. The yeast is already present on the fruits, so there is no need to make a wild yeast starter.

2. Place the sugar inside the jar and fill it with water. I like to leave ½ inch (1.25 cm) of headspace. Screw the lid on (not too tightly, though; you want fermentation gases to escape). Three times a day, screw the lid down tight and shake for 20 seconds or so, then loosen the lid again.

3. Taste after a week. If the yeast is very active, it will have converted most of the sugar to alcohol and your drink may get a bit tart and sour. Add a bit of sugar (¼ cup/50 g or so) and taste again in 2 or 3 days.

There are no rules telling you when you need to stop. You can stop the process when you like the flavors, or you can ferment it for quite a while, usually 3 to 4 weeks, and keep tasting and adding sugar as you go along. The longer you ferment it, the more alcoholic your drink will be.

When you're happy with the flavors, just strain the liquid into a smaller jar or bottle, then place it in the fridge to slow down the fermentation. Drink it within 1 to 3 days to keep the flavors from changing too much (see also the following section, "Pasteurizing Your Wine"). Be advised, however, that even in the fridge fermentation will continue, albeit at a slower pace. Monitor your container to make sure excessive pressure isn't building up; if it is, be sure to release it.

Pasteurizing Your Wine

"I love my wine! It's perfect—exactly the amount of sweetness (or dryness) that I wanted.

"But it's still fermenting! Damn yeast! Make it stop!"

Well, you can! Pasteurizing your wine is an option. It probably will change the flavor a tad, but it works nicely to stop a fermentation. Some people might think it's a crime to pasteurize a live fermentation, but you don't have to listen to them. It's *your* wine and you can do what you want with it. If it tastes great at the end, congratulations!

If you brew in 1-gallon (3.78 L) bottles, it's super easy to pasteurize. If you've made a larger quantity, you'll need to bottle it first and then pasteurize in the bottles. I like to pasteurize my wine while it's still in my 1-gallon fermenter, then pour the wine into my regular, thoroughly cleaned or sterilized wine bottles and cork them.

To pasteurize, place your bottle(s) in a large pot and pour in enough water to cover most of the bottle—as high as you can.

Heat the water until the temperature inside the bottle reaches 165°F (74°C) and hold it at this temperature for 10 seconds, then remove the bottle. Protect the top (some people place aluminum foil on top) until the

wine has cooled to room temperature, then cork. The method is based on the FDA-recommended pasteurization time and temperature for juices. I go overboard with 10 seconds (it's 2.8 seconds for juice).

Blending Your Wines

Last year I made some elderberry wine that ended up way too sweet, and recently I made some wild berry wine that ended up too dry. Wild yeast can sometimes play havoc with your recipes and change the rules to keep you on your toes.

I didn't know what to do with my berry wine, and I'd actually had forgotten all about the elderberry wine in the back of the pantry (usually I make vinegar with wines I'm not too fond of). As I was cleaning my shelves, I found both bottles and realized that they were the perfect partners for a blend. And were they ever! I simply mixed the two wines and bottled them. I enjoyed them 6 months later and it was one of the best wild berry wines I've made so far.

The moral of this story is, don't throw out a wine that may be too sweet or too dry; keep aging it and it may end up perfect for a tasty blend. Don't forget that you can also just add sugar to an already fermented wine that's too dry.

Burying Your Wines or Beers

Due to the cost of living in Los Angeles, I've spent the last two years living in a tiny house, which can get very hot during the summer. It's a real problem when I'm trying to preserve food and drinks.

One of my solutions is to visit my friend's place in the mountains (where it stays cooler during the summer) and bury my wine or beer bottles to age them. Temperature fluctuations are less extreme there than at my place, and I think there's something beautiful about aging a wine or beer that represents an environment by burying it in that exact environment.

Wine placed into the ground to ease temperature fluctuations while it ages.

I know it may seem quite odd, but if you research wines in the nation of Georgia, you'll find that this technique dates back several thousand years. There, wines are fermented and stored in large clay pots known as *kvevri* placed into the ground.

Making Mead (Honey Wine)

Mead is probably the easiest alcoholic fermentation. It's easy to imagine that the first mead could have been made by accident. Just a bit of honey left in a pot, a bit of rain, and in a couple of weeks someone had a delicious boozy beverage.

Pure raw honey contains the wild yeast necessary for the fermentation. The only problem is managing to get some actual pure, natural, *really* raw honey, which so far I haven't found at my local stores, even when the label says it's raw. I've come to the conclusion that most store-bought raw honeys and even the versions sold at farmer's markets (I found one exception) aren't really raw but have instead been adulterated somehow, because wild yeast is absent.

I'm not saying those honeys are bad, but if you want to do a truly wild fermentation using raw honey, you'll need someone who can provide you with straight-out-of-the-hive honey. With true raw honey, my success rate at starting a fermentation is pretty much 100 percent. I've used raw honey to make yeast starters for sodas, beers, wines, and of course mead-like drinks.

My method for simple meads is very basic, and so far it's worked well for me. I mix raw honey and water in a clean container and add my herbs. I let the plants macerate for 3 to 4 days, then strain and age the mead.

As with beer, some herbs (such as hops or mugwort), spices, or dried berries are much better boiled to extract the flavors. If that's what I'm working with, I'll boil my ingredients (water and plants/fruits), cool the liquid to lukewarm, then add my raw honey and stir to dissolve it. Next I strain the liquid into a clean container, then pitch the yeast. I use a raw honey yeast starter, usually from the same honey I used to make the mead, for the fermentation. If you have a potent and flavorful honey, you can also skip adding plants or other ingredients and let the original flavors shine.

How long should you ferment your mead? There are no hard-and-fast rules. You can enjoy your mead quite young or wait until it's fully fermented;

Wild berry wine corked and sealed with pinyon pine sap.

it's all about having the flavors you like. You will get flavors from the ingredients themselves, but your choice of honey will also influence the final result. Some of my meads were fantastic after about 1 to 3 weeks of fermentation but ended up kind of tasteless after a year. One of the best meads I ever made included lemon balm, and I aged it only for 2 weeks; it was still quite sweet, and the flavors were outstanding.

If you're really into making aged mead, you should read the book *Make Mead Like a Viking* by Jereme Zimmerman, which has excellent recipes. For my part, I tend to mostly make young herbal meads, although I have a couple that have been aging for more than three years.

How much honey should you use for making mead? In my book *The New Wildcrafted Cuisine* I stuck to the rule that, if you used wild yeast from raw honey, 1 pound of honey (a bit more than 1½ cups or 360 ml) mixed with 1 gallon of water (3.78 L) will give you around 5 percent alcohol.

Since I wrote that book, I've noticed that some wild yeasts can go higher than 5 percent. This was the case with raw honey I acquired while in Oregon and Vermont. You'll probably have to experiment a bit with your local wild yeast and terroir.

For fully fermented mead, I now tend to follow Jereme's suggestions and recipes to create basic, dry, or sugary meads, but I stay on the lower side for the amount of honey used because I strictly use wild yeast. For my regular herbal meads, which aren't fully fermented, the amount of honey used is just based on taste.

To find ingredients you can use to flavor your mead, check chapter 4's lists of plants, roots, spices, fruits, and so on used to brew beer; they all work for mead. Some quick online recipe research will also suggest ingredients you can find in your local terroir.

As a final note, there are *many* types of meads you can brew. From a mead purist's perspective, the traditional recipes above are technically not called mead but metheglin due to the addition of spices/herbs. Some other types of mead would include:

Melomel. Mead made with fruit.
Pyment. Grape wine with honey or mead made with grape juice.
Cyser. A mead made with apples or apple juice.
Show mead. Plain mead, the flavors coming from the honey itself (for instance, grapefruit flower honey).
Short mead. A quick fermentation with low amounts of honey.

BASIC TRADITIONAL RECIPES FOR AGED MEAD USING WILD YEAST

Dry to Semisweet Mead

1.5–2 pounds (0.7–0.9 kg) honey

1 gallon (3.78 L) water

¾ cup (180 ml) yeast starter

Spices and herbs of your choosing
   (optional—you can make a mead simply
   with honey and water)

10 or so raisins (I sometimes use sugary
   dried wild berries such as currants or
   blueberries), for added yeast nutrition

Juice of 1 lemon (optional, but good if you
   only used honey and water)

Sweet Mead

2½–3 pounds (1.14–1.36 kg) honey

1 gallon (3.78 L) water

¾ cup (180 ml) yeast starter

Spices and herbs of your choosing
   (optional—you can make a mead simply
   with honey and water)

10 or so raisins (I sometimes use sugary
   dried wild berries such as currants or
   blueberries), for added yeast nutrition

Juice of 1 lemon (optional, but good if you
   only used honey and water)

Procedure

Do a complete fermentation (which can take months) and transfer the mead to regular wine bottles, but make sure the fermentation is truly complete before doing so.

You can also make bubbly meads similar to sparkling wines by priming the mead when you bottle it (though difficult, make sure it's still very, very slightly fermenting). I've used 1.5 ounces (42 g) of additional honey with success (see "Civilized Carbonation in Bottles" on page 100).

Modern recipes, and some traditional ones, too, tend to add tannin and acid for better flavors. Acid isn't a must; more often it is used for very simple meads (just honey and water).

I don't add tannin, because a lot of the wild ingredients I use already contain some (yarrow, oak bark, and mugwort are good examples). If the plants, spices, and roots you use to flavor the mead don't contain tannin, Jereme Zimmerman suggests using one bag of black tea.

Commercial yeasts such as champagne or wine yeast can yield very high amounts of alcohol. I don't have experience using commercial yeasts with mead, and the recipes would need to be adjusted. That's where you need to do your own research or buy Jereme's book.

Primitive Herbal Meads

These are the meads I have the most experience with. I still call them mead because they fit within the definition:

MEAD:

An alcoholic drink of fermented honey and water.

But most of my fermented honey concoctions are aged for only a few days or weeks, and the level of alcohol can be very low. More often than not, I'm more interested in the flavors or medicinal virtues of the infused herbs and berries than the alcoholic content. As such, I monitor my ferments as they go along very much like I do my lazy wines.

Like wild beers, meads are a beautiful way to explore the flavors of whole environments (mountains, chaparral, desert, forest, and so on). You can also apply the same concept to your garden or use purchased plants/berries.

The photo on page 203 shows a forest blend I made for one of my fermenting workshops. It's composed of flavorful ingredients I chose to represent a specific forest and its seasonal qualities. The blend includes the following plants and herbs:

Willow leaves	Australian bush berries (dried)
Black sage	California sagebrush
Yerba santa	Fig leaves
Water mint	Mugwort
California everlasting	Lemons

It's made with a cold infusion process: The ingredients are placed in cold purified water with raw honey as the source of sugar and wild yeast. The jar can hold ½ gallon (1.89 L) of liquid. To start with, I use a bit more than a cup of raw honey for around ½ gallon, stir the liquid with a clean wooden spoon to make sure the honey is well diluted, and close the lid. Three times a day, I repeat the same stirring ritual. Within a couple of days—sometimes just 24 hours (in Southern California)—my fermentation starts. I strain the ingredients after 2 or 3 days and either bottle the contents as a soda, drink it as is, or continue fermenting it in a bottle fitted with an airlock or a similar container. If you keep fermenting it, keep tasting from time to time and add honey if necessary. There are really no rules; you can simply stop when you like it and place the liquid in the fridge, where the temperature will slow down the fermentation.

It's fun and so easy to do! The real skill is in creating the blend, and that's something that will only come with experience. My favorite blends are made with mountain ingredients such as pinyon pine, white fir, and crushed manzanita berries.

If you take a look at my concoctions, you'll note that I usually choose one main flavor profile (water mint in this example) and build everything around it.

Black sage, bush berries, and fig leaves have some nice sweetness, which is balanced by some bitterness from willow leaves (only five leaves were used), yerba santa (also aromatic), California sagebrush (a tiny amount), and mugwort. It's 70 percent mint, 20 percent sweetness, and 10 percent bitter. The end result is a sweet and extremely complex drink. I think it truly tastes like my forest in September.

Summer Forest Herbal Mead

FRESH MINT AND LEMON HERBAL MEAD

This is a good introduction to herbal mead-making: You just need mint and lemons. I usually use local wild mints (we have over seven species in Southern California), but you can also purchase some at your local store or grow your own. We'll look at a couple of methods to make it. This mead is best enjoyed young, and thus it's pleasantly sweet, but you are free to experiment with aging it as well. You need raw honey for this recipe, which makes around 1 gallon (3.78 L).

13 cups (around 3 L) water—don't use tap water, which may contain chlorine
2–3 cups (0.45–0.7 L) raw honey
2–3 large bunches mint
4 lemons

Procedure

Before you begin, make sure you thoroughly clean everything you're going to use during this procedure. This includes the ingredients.

1. Mix the water and raw honey in a large jar or plastic food-grade container. Stir until the honey is dissolved.
2. Roughly chop the mint and cut the lemons in wedges. Place everything in the water/honey solution. Stir one more time to make sure everything is well mixed, then cover with a clean paper towel or cloth tightly secured with a rubber band or string.
3. Three or four times a day, remove the cover, take a clean spoon, and stir everything for 10 to 20 seconds. This helps activate the wild yeast in the solution and, in my experience, speeds up the fermentation and avoids potential spoiling.

Before using it, I clean the spoon with regular soap and water, the idea being to remove any unwanted bacteria. After stirring, cover the top again.

4. Depending on the temperature and the willingness of the local fermentation gods, after 3 to 5 days you should see some bubbling. This means the fermentation is active. As an added insurance, you can smell it—a good fermentation does not smell bad.

   Another indication that the fermentation is quite active is the fact that your ingredients (mint and lemon rings) will be floating on the surface because of the large number of bubbles in the mix. I like to remove and strain my ingredients after 2 to 3 days.
5. Using a clean sieve and funnel, your next step is to strain the solution into a 1-gallon (3.78 L) jug.

   Again, make sure you clean your sieve and funnel properly so you don't introduce unwanted bacteria into your mead. I use regular dish soap with hot water and, so far, have never experienced any problems.
6. Place an airlock on your container or cover it with a clean paper towel or cloth tightly secured with a rubber band or string. Wait 2 to 3 weeks before enjoying.

Step 2

Step 3

Step 4

Step 5

Step 6

BOILING TO EXTRACT FLAVORS

For some wild ingredients such as mugwort, hops, yarrow, mushrooms, and various roots, boiling may be necessary to extract the flavors.

Procedure

1. Boil your ingredients in a pot for 15 to 20 minutes. I usually place 50 percent of my raw honey with the ingredients. When done, let the solution cool by placing the hot pot in cold water to speed up the cooling process (adding ice to the water is a good idea). To protect the brew from unwanted bacteria, keep the lid on while cooling the pot.
2. Once the solution is lukewarm (70°F/21°C), pour it, with or without straining it, into a glass or plastic food-grade container, then add fresh herbs and the rest of your raw honey (which contains the wild yeast)— usually I use 2 to 3 cups (500–700 ml) of honey per gallon (3.78 L) for this type of brew. Stir until the honey is dissolved, then cover with a clean paper towel, kitchen towel, or cheesecloth. Use a string to tie it up nicely so flies or unwanted bacteria don't get in.
3. Three or four times a day, remove the cover and use a clean spoon to stir everything for 10 to 20 seconds. Depending on the ambient temperature, after 3 to 5 days you should see some bubbling. This means the fermentation is active.
4. Using a sieve and funnel, your next step is to strain the solution into your fermenter (bottle or similar container).
5. Place an airlock on the container or cover it with a clean paper towel or cloth tightly secured with a rubber band or string. Wait 3 weeks before enjoying.

No Raw Honey?

If you don't have raw honey, just use the same amount of honey asked for in a recipe and add around ½ to ¾ cup (120–180 ml) of wild yeast starter (ginger bug or the like) per gallon (3.78 L). Then continue with the usual procedure.

Turning Your Hike into Mead

Ipsa scientia potestas est—Knowledge itself is power.

When Francis Bacon wrote these words in *Meditationes Sacrae* (*Sacred Meditations*), he probably wasn't referring to nature or fermentation. As with many universal truths, however, I think these words are extremely relevant to what I do.

My sacred meditation is often in the morning and in the form of a hike. I sometimes call it my sacred hike, but it's really more a slow walk. The goal is not to exercise my body, and in fact I make it a point not to have a goal. I just let my mind and senses soak in the smell, sounds, texture, and beauty of nature. As I said in the beginning of this book, it's extremely Zen.

You need to let go of all the noise in your head and simply *be* there. It may not be easy for the first few minutes, and it definitely helps if you didn't read or listen to the usual dreadful morning news that day, but at some point it just happens naturally and you become one with nature. It's very much like turning on a switch, and if you do it several times during the week, you'll find it gets easier with time.

It's quite a euphoric feeling and extremely spiritual. It's as if your consciousness becomes your surroundings: You are the trees, the stones, that bird singing, the flow and sound of the stream; you are one with the forest itself. The simplicity of it is hard to explain—it's an experience. As humans, we've had such a deep connection with nature for millennia that it's part of our DNA, and without it, it's like an important piece of ourselves is missing. I know I get a bit depressed or moody if I don't get my "dose" of nature.

I wish I could stay in that euphoric state forever, but, speaking of Zen, there is also the famous proverb: "After enlightenment, the laundry." Eventually you have to go back to your day-to-day world and deal with its imperfections—doing the laundry, paying the bills, having a job, making money, and so on . . .

But I'm rambling a bit. To return to the subject, the hike mead came about one particularly boring day, which thankfully doesn't happen much anymore. I was working on retouching some photos for a client and decided to take a sip from one of my mountain sodas, which I had brewed a couple of weeks earlier. Just as it happens during my sacred hikes, the switch got turned on and, with every delicious sip, I was transported back again to the locations where I'd collected the ingredients. It was quite amazing, like being transported back in time and re-experiencing the sights, emotions, thoughts, and feelings from my original trip.

A forest hike mead. Notice the change of color due to the wild cherries.

It helped that the work I was doing at home in the first place was so dull. Still, this moment really inspired me. I realized that you can preserve food through various processes like fermentation, but you can also preserve feelings and moments in time to re-experience later on. It's much like looking at a photo of someone you love or the photo you took of a beautiful location while vacationing. Looking at it can bring back emotions and a smile.

And so I became fascinated with the notion of bottling not only whole environments like the mountains, forests, or chaparral but also specific (smaller) locations and emotional moments. Anyone can do this, and it doesn't even have to be a hike in the wilderness. You can make a drink out of a walk in your garden or a summer morning stroll in your neighborhood.

Your only limitation is knowledge, and that's where "knowledge itself is power" comes into play. It's well worth investing your time and money to learn about your local plants through books, online research, and most important connecting with people who have knowledge and experience. I learned a lot from elders while growing up in Belgium and also locally.

Had I made a drink from a hike 20 years ago, I probably would have ended up with a simple mint mead. With the knowledge I have accumulated since then, I can choose from 60 or more possible ingredients during a 30-minute walk in most environments.

The more knowledge you accumulate about your own local plants, the more creative power you'll be granted by nature's spirits. It's really infinite, and it doesn't matter where you live—you can create amazing drinks with your immediate terroir.

So where are the recipes? The techniques?

It's Zen! The recipe is really just water, raw honey, and that hike in a jar. The ingredients nature and inspiration provided that day. It can be simple or extremely complex; it's a reflection of that walk and that moment in

Forest hike mead: raw honey, willow and fig leaves, wild cherries, rabbit tobacco, grass, and yarrow.

time. It changes all the time. I could take the exact same hike each day and create something different with it.

It's a fun approach. I know I'm going to create something out of the walk/hike, but instead of trying to figure out what in advance, I just let nature inspire me and dictate what the drink will be.

You can just *be* there and let the drink happen. It's something you can do when you have accumulated enough knowledge and experience; you simply let it all go and be inspired.

The technique is also very loose. I clean my ingredients briefly and place them in a jar, then add water and raw honey to taste. It just needs to taste sweet to be fermentable. I often start with ¾ to 1 cup (180–250 ml) of honey for ½ gallon (1.89 L) of water. A smaller amount of raw honey will make the drink turn sour quite fast, and a larger amount can make the drink sweeter, or more boozy if you ferment it for a while. You can even add more honey or sugar as you go along, and when you're happy, drink it as is or place it in the fridge to slow down the fermentation.

You can also carbonate it if you want by placing your hike mead in a swing-top or recycled soda bottle.

In the beginning, you may have some hikes that are so-so in terms of flavors, but very quickly you'll be able to create drinkable masterpieces. As usual, the fun is in experimenting.

Be Zen and have fun!

CHAPTER 8

Ethnic Drinks and Medicinal Brews

Doing some research on local ethnic drinks is always a worthwhile activity and an interesting way to get started brewing with ingredients from your own native garden or the wilderness surrounding you. Such drinks are a perfect example of people brewing with whatever their immediate environment provides, and are often a true representation of terroir brewing. I also learned a lot about sources and extraction of wild yeast by studying this subject.

Many ethnic drinks don't fit the usual classification of alcoholic beverages into as wines, beers, liqueurs, and so on. The saguaro fruit wine made by the Tohono O'odham people could be classified as a wine, but some fermented drinks like *cauim*, an alcoholic beverage made of cassava root, or *lubisi*, made with banana and cereals, can be a bit difficult to classify. *Chicha*, a fermented beverage in which saliva is used to convert maize starch to sugar, could be called a beer, but I'm not sure that classifying all these drinks is that important. They are what they are, beautiful in their own ways.

Such drinks are often really simple, many using open fermentation and wild yeast, but you do get exceptions. Some recipes are extremely easy to make while others use a wide variety of ingredients such as fruits, grains, berries, barks, aromatic plants, and so on.

There are probably hundreds, if not thousands, of these interesting regional drinks in the world. The book *Ethnic Fermented Foods and Alcoholic Beverages of Asia*, edited by Jyoti Prakash Tamang, lists more than 50 such traditional alcoholic drinks in India alone.

Over the centuries, due to human migrations, modernization, and the ability to import edible goods from more distant lands, some of these drinks have adapted. A good example is the pre-Columbian drink tepache. The word comes from *tepiatl* (in the Náhuatl language), meaning "drink corn," and as the name indicates it was originally a traditional drink made with corn in central Mexico. The original recipe was altered when pineapple was introduced. Now tepache has become a fermented pineapple drink with added raw sugar and spices.

Sometimes you come up with new beverages by using traditional recipes with local ingredients. A good example is smreka, the slightly alcoholic drink from Bosnia made by fermenting juniper berries. I already used my local California juniper berries as a source of wild yeast, but never thought about doing a full fermentation with them. Due to the fact that my local berries have a completely different flavor profile (more lemon and pine flavors) than the ones from the common juniper (*Juniperus communis*), I end up with a similar beverage, but the flavor is completely local.

In my area (Los Angeles), there is little to no evidence of alcohol production from the original inhabitants, but for inspiration I can look at traditional drinks made nearby, such as in Arizona or Mexico; we share a lot of the same fruits and plants. For example, I found out recently that, in Mexico, a fermented drink was made using the berries of the Peruvian peppertree (*Schinus molle*). The recipe seems lost to history, but it would be interesting to experiment a bit—the trees are very abundant locally.

So research your own local history online or through books, and talk to local elders about possible fermented drinks they recall. You'll be amazed at what you can find.

TEPACHE

Tepache is a well-known traditional fermented Mexican drink made with the tops and peels of pineapple. In Los Angeles I sometimes find food carts selling the drink and serving it cold in a small plastic bag with a straw placed in the center. It's super refreshing during the summer heat.

The recipe is simple: just pieces of pineapple with raw brown sugar (piloncillo), cinnamon, and cloves. Of course, you will find variations here and there; some people like to use the whole pineapple for added flavors, not just the tops and peels. Other spices can also be added, including anise seeds, orange peel, cardamom, allspice, and even dehydrated chile pods for a spicy kick.

In late spring I can forage an interesting flavorful plant locally called pineapple weed (*Matricaria discoidea*). The flowers taste like a cross between chamomile and pineapple. Last year I made a "Southern California Tepache" using these flowers instead of pineapple, with excellent results.

You have a lot of room for creativity! Very interesting and tasty drinks can be made by substituting other fruits for the pineapple or simply adding them to the traditional recipe.

1 ripe pineapple

3–4 quarts (3–4 L) water

1–2 piloncillo cones (8 ounces/224 g each); you can substitute brown sugar or even maple syrup

1 stick cinnamon

3–4 whole cloves

1 dehydrated chile pepper pod (optional, but I like a spicy kick)

Wild yeast (this comes from the pineapple skin)

Procedure

1. Chop the pineapple (you can just use the skin and core, or use the whole fruit for added flavors). Combine all the ingredients in a large container (glass jar, clay pot, or food-grade bucket).

2. Cover with a clean towel and stir three times a day until you see some nice fermentation going. Strain when the fermentation is doing well (usually 2 to 3 days).

   There are no real rules as to fermentation time. This brew is usually drunk right away, but you can ferment it for a few more days to get a higher level of alcohol (some people even add beer to it). Don't wait too long, though, as eventually it will turn to vinegar.

SMREKA

I originally discovered this drink while reading the book *The Art of Fermentation* by fermentation guru Sandor Katz. I immediately became interested in this recipe, as we have tons of edible juniper berries locally.

Smreka is a lightly fermented drink made in Bosnia by placing juniper berries in springwater (don't use tap water, which may contain chlorine) and letting nature do its thing. The berries contain some sugar and are loaded with wild yeast. Add some water and you end up with a very slow fermentation. Making smreka is bit similar to making a lazy wine, but it's even easier. There is nothing to do aside from shaking the bottle once or twice a day until you get a fermentation going (you'll see the usual small bubbles); after that, shake maybe once or twice a week. Initially the berries float on top, infusing their flavor into the water. As the fermentation goes on, the berries will begin dropping downward, and the fermentation is complete when all the berries have sunk to the bottom.

Before you get too excited, although juniper trees can be found in many countries, realize that not all juniper berries are edible, so you need to do your homework before making your own local smreka. Experimenting with my local berries, I found out that my drink tasted much better if I used somewhat unripe berries, which have some wonderful pine/lemon flavors. The very ripe berries don't have too much flavor and are extremely sugary. Smreka made with the ripe berries ended up quite sour and flavorless, while a batch made with the unripe berries was delightful. You may need to experiment with your local berries.

2 cups (500 ml) juniper berries
1 gallon (3.78 L) springwater
Sugar or honey (optional)

Procedure

1. In a gallon bottle (or jar), combine all your ingredients. Place an airlock or clean towel on top. If you used a jar, don't screw the lid too tight; you want fermentation gases to escape.
2. Two or three times a day, shake or stir your fermenting vessel. After 3 or 4 days (sooner during summer), you should see some bubbling going on, which is an indication that your fermentation is doing well. At this point you can simply let the fermentation continue until all the berries have sunk to the bottom;

this can take anywhere between 2 weeks and a month depending on the temperature and how active the fermentation is. I like to shake the bottles once or twice a week when the fermentation is happening.

3. When the fermentation is done, strain the liquid into a bottle and place it in the fridge. It will last for at least a month, if not longer.

Notes: *It's completely okay to add a bit of sugar while making smreka—usually white sugar, but why not try using honey or maple syrup? This will make the drink a bit more alcoholic. You can also judge by flavor; taste as you go along and stop when you like it. The last time I made it, I fermented a bit too long and it ended up sour, even with the unripe berries. Adding a source of sugar at the end to make the drink sweeter is an option.*

You can also make a soda by adding sugar to your smreka and siphoning the liquid into recycled soda or swing-top bottles (see "Wild Carbonation in Bottles" on page 99).

TISWIN (AKA COLONCHE)

Tiswin is an alcoholic drink brewed from corn, but it's also the name for the sacred drink made by the Tohono O'odham people, who reside in the Sonoran Desert of Arizona and Mexico. From my research, it seems that the traditional wine used raw fermentation, but more modern recipes boil the juice, probably because it would keep longer. I've made both versions using my local prickly pears; interestingly enough, boiling the juice changes the flavor profile, making it taste a bit like corn. I like the raw fermentation much better, but both are worth making. This recipe makes around 5½ cups (1.3 L).

Around 30 large nopal cactus fruits (or 60 prickly pear fruits), cleaned and needles removed (see instructions in Lazy Prickly Pear Wine recipe, page 194)

3 cups (0.75 L) water—don't use tap water, which may contain chlorine

Yeast (wild yeast starter or packaged dried champagne yeast)

Procedure

1. Remove the skins from the cactus pear fruits and set aside the pulp. It's actually quite easy: I usually cut the fruits in half and scoop out the insides using a spoon. You can also hold the sides of the halved fruits with both hands and with your thumbs push in the center to remove the insides.

2. Place the pulp in a blender and add the water. Blend at low speed for 5 seconds or so, then pour the raw juice into a strainer. You can also use your clean hands or other instruments if you don't have a blender. Once the mix is in the strainer, I use a wooden spoon to stir it vigorously and push the juice through the mesh. It usually takes a few minutes to separate the seeds from the juice.
3. Pour the juice into a large pot with no lid and bring the liquid to a boil, then lower the heat and cook at a slow simmer for around an hour. Watch carefully when the liquid is coming to a boil; there is a substantial amount of foaming, and it can easily spill over. Trust me, I know!
4. Strain the liquid again. You may notice that the color has changed from a dark red to a beautiful orange. This time be gentle and don't force the pulp through the mesh; you want a nice golden liquid without any impurities. Return the liquid to the stove for another 20 to 30 minutes of simmering, which will concentrate the sugar even more through evaporation.

By the way, there is little consensus as how much time the initial and second boilings should take; it can range from 20 minutes to 2 hours. I've found 2 hours to be excessive, and I had very little liquid left in my pot at the end, which didn't make sense. The longer time was probably necessary when making a large quantity of tiswin. The timing in this recipe is based on my own trial and error using a fairly small amount of cactus pears.

5. Remove the pot from the heat and place it in cold water (with the lid on) until lukewarm (70°F/21°C). You may need to change the cold water several times and add ice to the water to speed the process.
6. Once lukewarm, add the yeast (½–¾ cup/ 120–180 ml for wild yeast starter). Pour the liquid into a bottle and fit it with an airlock. My research tells me that, like many primitive drinks, this beverage was enjoyed before it was fully fermented, usually in 4 to 5 days, but you can also do a full fermentation if you want.

MANZANITA FERMENTED CIDER

Cider has two definitions. The first one is "An unfermented drink made by crushing fruit, typically apples." The second is exactly the same, but now the drink is fermented and alcoholic. So you can make both and still call them both ciders.

Manzanita is perfect for making cider; in fact the name itself is the clue, as *manzanita* means "little apple" in Spanish. The ripe dry berries usually taste like sweet lemony apples, and the unripe green berries have a flavor similar to sour green apples. Of course there are exceptions; bigberry manzanita (*Arctostaphylos glauca*), for instance, has a skin that tastes like lemons and thus can be used as a lemon substitute in some recipes.

The plant can be found in the Southwest from Texas to California as well as in Mexico. Some species of manzanita are rare and endangered, so take the time to learn about them and where those can be found. By doing so, I learned that in my area there are no endangered plants. Renowned forager, hunter, and friend Hank Shaw has good advice: Only pick from large masses of the plants, not isolated individuals.

I usually do my collecting around August or September, when the berries are fully ripe and dry. They are extremely plentiful in the local mountains; one friend has a whole section of her property loaded with them. In one morning I can collect a large grocery bag.

4–5 cups (1 L) dry manzanita berries
1¼ pounds (567 g) light brown sugar
1 gallon (3.78 L) water
Yeast (beer yeast or wild yeast)

Procedure

1. In order to extract the flavors, you need to crush the berries. I use my *molcajete* (stone grinder). I don't think it's a good idea to do this in a blender, as the seeds are very hard and could damage the blades. It's much better if you go primitive and find some rocks to crush them. I like to remove as many seeds as possible, since most of the flavors are in the dry skin and flesh.
2. Once you've ground the berries to the consistency of a coarse meal, place them, along with the sugar and water, into your pot. Boil for 30 minutes.
3. Cool to 70°F (21°C), then add the yeast (½ cup/120 ml for wild yeast starter). Place the liquid into the fermenter, top it with an airlock, and let it ferment for 10 days. Start counting when the

fermentation is active (this may take 2 to 3 days with a wild yeast starter).

4. Clean your funnel/sieve and strain the liquid into 8-ounce (237 ml) swing-top bottles. Prime each bottle with ½ teaspoon (2 g) brown or white sugar for carbonation. (I usually use white sugar.) Close the bottles and store in a place that's not too hot. The cider will be ready to drink in 3 to 4 weeks.

Note: You could use more sugar with commercial yeast, do a full fermentation, and enjoy a nice non-carbonated but quite alcoholic drink.

Break the bread slices into small pieces. Place them in a preheated 350°F (177°C) oven for 10 minutes, then broil on high until golden brown.

Pour the water and sugar into a pot and bring the liquid to a boil. Place the toasted bread, mint, and raisins into the boiling liquid and stir briefly. Bring back to a boil, then remove from the heat.

Place the pot (with the lid on) in cold water and cool the liquid to 70°F (21°C), then add the yeast.

Ferment for 8 to 12 hours, then strain the liquid into a bottle or into recycled soda bottles if you want carbonation.

TRADITIONAL BREAD KVASS

My grandpa was very, very Belgian and loved his beers. I remember a time we visited him when I was around five years old. It was still early in the morning, and there he was, sitting at the table, having his favorite breakfast. It was a simple one consisting of long strips of sliced bread that he dipped into beer and ate with absolute relish. Watching him, I could not make up my mind: Either I was in the presence of a genius or it was one the most unappealing breakfasts I'd ever seen.

I was born with an inquiring mind, probably my curse, and thus I asked him if I could try a bit of this possibly awesome and innovative dish. He handed me a piece after dipping it into the beer, and upon placing it in my mouth and chewing on it, I had the revelation that my grandpa wasn't a genius. I still loved him after that, but gosh, it was definitely an acquired taste. Maybe it's that childhood trauma that caused me to avoid making kvass until very recently. In my head, bread and a fermented drink didn't seem to fit together.

From a historical perspective, though, I could not have been more wrong. Bread and beer have a history together that goes back thousands of years, to when bread was one of the main ingredient in some Egyptian beers. In fact it's probable that bread and beer were invented around the same time. Both share the exact same ingredients: grains, wild yeast, and water. Some breads, like beers, use herbs as flavoring agents.

Kvass is an interesting Slavic/Baltic drink, and probably comes from the same heritage as those ancient drinks. It's not really a beer but a fermented soda-like beverage made from bread (itself made from barley, wheat, rye, and so on). It's often flavored with fruits, berries, or herbs. It's usually not very alcoholic, though you can find some recipes for strongly alcoholic kvass.

Like those ancient beers, kvass was mostly a "people's" drink, similar to the weak Belgian *saison* beers that were meant to quench the thirst of the working classes while adding some valuable nutrition to their diet. Due to its bread content, the drink is a good source of vitamins and calories. It is often advertised as a drink to promote digestion and a healthy gut.

Today kvass is very much enjoyed like a soft drink and often carbonated. Like commercial sodas, however, the modern fizzy versions sold at the stores are often a far cry from the original recipes and use corn syrup, sugar, malt extract, and artificial flavorings.

But we can keep the tradition of making healthy drinks alive! Kvass is easy to make and quite enjoyable, and despite my childhood trauma I now like it very much. Changing the ingredients allows you to make countless types of kvass: wild currant kvass, mint kvass, and so on. My local Middle Eastern supermarket even sells kvass flavored with thyme.

½–1 pound (227–454 g) rye (or other) bread

1 gallon (3.78 L) water

1½ cups (300 g) sugar (I like to use brown sugar, but honey is okay, too)

Small handful of dried herb(s), for flavoring (usually dried mint, but I've seen basil, thyme, or rosemary used)

0.4 ounce (11 g) raisins

½–¾ cup (120–180 ml) wild yeast starter or commercial beer yeast

Procedure

1. Slice the bread and break the slices into smaller pieces. Place them in a preheated 350°F (177°C) oven for 10 minutes, then broil on high until golden brown—this can take 3 to 5 minutes depending on your oven. You basically want your bread to look like slightly overdone toast, which will give the brew a nice amber look and better flavors.

2. Meanwhile, pour the water and sugar into a pot and bring the liquid to a boil. Place the toasted bread, herbs, and raisins in the boiling liquid and stir briefly. Bring the liquid back to a boil, then remove the pot from the heat.

3. Place the pot (with the lid on) in cold water and cool the liquid to 70°F (21°C), then add the yeast (½ to ¾ cup or 120–180 ml for wild yeast starter).

   Either keep everything in the original pot with the lid on or transfer the contents into a fermenting bucket fitted with an airlock or a clay pot/glass container with a clean towel on top.

4. Ferment for 8 to 12 hours, until you see some bubbling going on, then strain the liquid into a bottle or into recycled soda bottles if you want carbonation. Check the pressure (see page 99) and place the bottles in the refrigerator when ready. Drink within a couple of weeks.

SOUTHERN CALIFORNIA KVASS

Of course, once you realize that kvass is a beverage that just requires bread, sugar, wild yeast (you can also use beer yeast), and water as the base—the rest is flavoring—the gate to making your own creative drinks swings wide open.

Instead of using dried mint or basil, you can use herbs from your garden or forage some flavorful wild ones. For my part, I also have tons of wild berries I can add to the drink.

Like my wild beers, I can easily brew some kvass representing all kinds of environments, such as the mountains or my local forest. Here is a recipe that I have made and enjoyed very much.

½–1 pound (227–454 g) rye (or other) bread
1 gallon (3.78 L) water
1–1½ cups (300 g) piloncillo sugar
0.2 ounce (6 g) local wild mint or other aromatic herb such as water mint (*Mentha aquatica*)
0.1 ounce (3 g) bitter herb (mugwort, yarrow, or California sagebrush)
3 lemons
1.2 ounces (34 g) or more local wild berries such as currants, coffee berries, or Mexican elderberries (mixing them is okay, too)
1 ounce (28 g) manzanita berries
½–¾ cup (120–180 ml) wild yeast or commercial beer yeast

Procedure

Use the same brewing method as for Traditional Bread Kvass, page 221. The lemons should be squeezed and thrown into the water with the sugar before boiling.

NORTHEASTERN KVASS

This recipe is based on the forest I like to hike in Vermont. It's a mix of pine and root flavors, a bit like a kvass root beer. It's quite enjoyable and nutritious. The method is a bit different, as the pine branches and spruce are not boiled. Of course, maple syrup is the source of sugar for this fermentation, and the wild yeast is from a dandelion flower starter.

½–1 pound (227–454 g) rye (or other) bread

1 gallon (3.78 L) water

1–1½ cups (250–360 ml) maple syrup

2 tablespoons (10 g) sassafras root bark

1 tablespoon (5 g) sarsaparilla roots (optional)

1 tablespoon (5 g) chopped dandelion roots

½ teaspoon (1 g) dried wintergreen leaves

Handful of turkey tail mushrooms
   (just because I like them and they're
   good for you)

Small piece (¾–1 inch/2 cm)
   gingerroot (optional)

A couple of small spruce or white pine
   branches, or any lemony-tasting pine
   needles (you can also use a couple of
   lemons if you want; juice them and then
   throw them in the pot)

½–¾ cup (120–180 ml) wild yeast or commercial beer yeast

Procedure

Use a similar brewing method as Traditional Bread Kvass (page 221). The main differences are that you can place the turkey tail mushrooms in the water from the start (at the same time as the sugar) so they boil longer than the other ingredients. The spruce or white pine branches are added when the liquid is cooled down and the yeast goes in. It's a personal choice, but I don't like the flavor of boiled spruce/fir/pine. Don't forget to cut the top of the needles so the flavors can be extracted.

Because I use lots of barks, dried leaves, and roots in this recipe, I don't place the pot in cold water but simply set it outside. The warm water cools slowly, and I extract more flavors that way.

FRUIT KVASS

As you've just read, kvass is a fermented beverage usually made with bread and often flavored with fruits, but there are a few other variations such as beet or fruit kvass. These probably developed as extensions of original recipes, but the bread was omitted, possibly due to dietary restrictions. And if you remove the bread from a regular kvass, you're basically fermenting fruits.

I think it's stretching the definition of a little bit, but if you do some research, you will find that, yes, there are such drinks as kvass made with just one fruit/berry or mixed fruits/berries.

Interestingly, some recipes use yeast (usually from a ginger bug) and sugar, while others recipes use whey as starters. Both methods work, but with the whey method (lacto-fermentation), the flavors are a bit more sour.

Procedure for Wild Yeast

1. Cut your fruits in fairly large pieces. Some berries may need to be smashed a bit.
2. Pack a ½-gallon (2 L) jar with enough berries and fruits to fill 70 to 80 percent of its volume.
3. Pour in ½ cup (120 ml) ginger bug (see page 48).
4. Add filtered water to almost fill the jar, leaving about a 1-inch (2 cm) headspace.
5. Add ⅓ to ½ cup (75–100 g) white sugar and shake the jar.
6. Don't screw on the lid too tightly; you want fermentation gases to escape. Three times a day, screw the lid down tight and shake for 10 seconds or so, then unscrew the lid again a bit. Depending on the temperature, after 2 or 3 days you should have a nice fermentation going. The drink is now ready to enjoy. It may be a good idea to place a plate under the jar, as sometimes when the fermentation is very active it can push the fruits/berries

up and some leakage may occur. This usually doesn't happen with large chunks of fruit.

Note: *You could also use ½ cup (120 ml) of raw unpasteurized honey instead of sugar. The raw honey contains wild yeast, so it's not necessary to use ginger bug.*

Procedure for Whey

1. Cut your fruits in fairly large pieces. Some berries may need to be smashed a bit.
2. Pack a ½-gallon (2 L) jar with enough berries and fruits to fill 70 to 80 percent of its volume.
3. Pour in ½ cup (120 ml) whey.
4. Add filtered water to almost fill the jar, leaving about a 1-inch (2 cm) headspace.
5. Add ⅓ to ½ cup (75–100 g) white sugar and 1 teaspoon (5.5 g) salt. Shake the jar.

   Finish your kvass using the same method as for fruit/berry kvass using wild yeast, above.

Fruit kvass using wild yeast, organic fruits, and locally foraged flavors (white fir and pinyon pine).

Adding Local Flavors

This is where the fun is. Sure, you can ferment fruit kvass using just fruits, but you can also create unique flavors by using local aromatic or flavorful ingredients. In many of my Southern California recipes, I use pinyon pine or white fir branches with the tips of the needles cut off so the flavors can infuse. This adds some citrus/pine qualities to the drink. I've made other kvass using my regular bitter herbs (yerba santa, yarrow, mugwort, California sagebrush), wild mints, or local aromatic sages (white or black sage).

If I lived in the Northeast, I would probably use local fruits and berries such as apples, pears, blackberries, raspberries, dewberries, and so on. To add more local flavors, I might add a white pine branch or spruce tips. Or I could push the flavors a bit toward a sort of root kvass by using wintergreen leaves or sarsaparilla root. Of course, maple or birch syrup would be the sugar source.

WILD BERRY–MINT
FERMENTED KOMPOT

Kompot is nearly the same as fruit kvass, the main difference being that kompot is really a food preservation technique to start with, while fruit kvass usually uses fresh seasonal ingredients. Then again, if you research the subject, you'll get thoroughly confused, as some people call kompot a fruit kvass and vice versa. Or maybe I'm the one who's confused.

But let's add even more confusion. You may have heard of "compote" before. Well, compote and kompot are also two different things. In the culinary world, compote is fruit stewed or cooked in a syrup, usually served as a dessert. It's not really a liquid; it's quite thick, like a purée, and sometimes you have pieces of fruits in it. In Belgium, my mom used to make a rough chunky compote with apples, brown sugar, and a tad of cinnamon. It was served as a side dish (not a dessert). The thing that compote and kompot have in common (hence perhaps the similarity of the words) is the fact that the fruits or berries are boiled.

So what is kompot?

It's very similar to fruits preserved in syrup, although in this case people are more interested in the flavorful syrup than the fruits. The fruits or berries are really there to create taste.

To keep it simple, kompot is a very sugary beverage made by boiling fruits and/or berries; sometime spices or herbs are added for flavor. Unlike fruit kvass, which uses mostly fresh seasonal fruits, kompot isn't so fussy— everything goes in. No rules! You can use fresh, dried, canned, or even frozen ingredients. No wonder kompot sounds like compost!

The boiling method makes sense, too, because some berries will unleash more flavors when they're dehydrated first. It's certainly been my experience with local elderberries and other dried berries such as currants or local toyon berries.

When you drink traditional kompot (unfermented), you need to add water to the syrup (otherwise it's just too sugary), but because we're fermenting it, we're going to add the water at the beginning. Excessive sugar isn't good for fermentation, since it acts as a preservative and actually inhibits the yeast from doing its valuable work.

As you can imagine, there are countless combinations of flavors you can experiment with and ferment. Just search "kompot" online; you'll find tons of recipes.

It's extremely important to have high-quality ingredients. The following recipe can be pretty much tasteless with some farmed blueberries, yet highly flavorful with organic or foraged blueberries picked at just the right time. Taste your concoction after 10 minutes or so, and you can always improvise by adding other interesting flavorful ingredients such as fennel seeds, sassafras, sages, and so on.

1–1½ cups (250 ml) blueberries (or cherries, cranberries, strawberries. elderberries, wild currants, sliced pears, etc.)

2 large sprigs mint (wild or garden variety)

2 quarts (2 L) water

½ cup (100 g) sugar

2 lemons

½ cup (120 ml) wild yeast starter or commercial wine/champagne yeast

Procedure

1. Wash the berries and mint, then set aside.
2. Pour the water into a pot, squeeze the lemons then throw them in the liquid, and add the blueberries and sugar.
3. Bring the water to a boil, then lower the temperature and simmer for 20 minutes.
4. Remove the pot from the heat, add the mint, and place the pot (with the lid on) in cold water until lukewarm (70°F/21°C). You may need to change the cold water several times to cool it faster. I usually add ice to the water to speed up the process.
5. Once the brew is lukewarm, add the yeast and pour the liquid (and berries) into a bottle. Fit it with an airlock. To make a soda, ferment for around 10 hours, then strain and place into a recycled soda bottle (or glass swing-top bottles). Check the pressure after a day and, if it's adequate, place the bottle into the fridge.

   Otherwise just ferment the kompot completely and drink within 2 or 3 days, like a primitive wine. You can also use the fermented berries in desserts.

Note: You can also place some fresh mint with the berries in the fermenter. Just bruise it a little so the flavors will be extracted in the cold fermentation.

Medicinal Brews

A whole book could be written on the subject of healing beers, sodas, or wines and their potential in a medicinal or health-based cuisine. A full treatment of the subject is beyond the scope of this book, and requires more specific knowledge than the average person making beers or sodas for flavor is likely to have. Still, if you're interested in herbalism and the medicinal uses of what nature has to offer, it's a beautiful and worthwhile territory to explore.

I highly recommend Stephen Harrod Buhner's book *Sacred and Herbal Healing Beers* if you intend to explore that path. It's a fascinating and informative read.

The more I delve into wild fermentation, the more I tend to think that, originally, flavors and medicine were not separate subjects. Our ancestors knew much more about plants than we give them credit for. Even today, knowledge of medicinal plant use is still plainly visible in many countries. When I purchased some dried curly dock (*Rumex crispus*) at my local Middle Eastern store in Los Angeles, under the APPROVED BY THE FDA English label I found the original label in Arabic and French. Half of the content described the medicinal virtues of the plant.

Sadly, in many parts of the world and particularly in Europe and North America, the old knowledge and oral traditions have been lost through the various conquests, invasions, and genocides so common throughout human history. We have paid a dear price for modern civilization. Two thousand years ago the Pax Romana effectively erased most of the Druidic oral traditions and medicinal knowledge. The hunt for witches and their subsequent burning at the stake in the name of religion also took a toll on traditional herbalism, as many of the so-called witch brews were simply medicinal drinks and potions. More recently, a similar genocidal process occurred in the United States and locally with Southern California natives. A lot of precious knowledge was lost again.

Interestingly enough, if you dig a bit deeper you'll learn that many of our modern drinks had medicinal origins. Even the modern sodas found at the grocery stores often have a healing ancestry. In the 19th century a drink like Coca-Cola was advertised as a health drink, "containing the properties of the wonderful Coca plant and the famous Cola nuts." Dr Pepper with its secret ingredients was marketed as a brain tonic and energizing drink. I'm not saying these drinks were particularly healthy to start with, but I'm sure that some were much better for us than the combination of cheap corn syrup, artificial flavors, and pumped CO_2 they now call sodas or "carbonated drinks."

On the plus side, there has been a definite resurgence recently in the study of herbalism and folk remedies. I think it's a worthwhile area of research, and the knowledge can now be written down and preserved through books and digital media, making it much more difficult to obliterate.

Ingredients for the chaga beer (page 236): mugwort, willow bark, chaga mushroom, yarrow flowers.

Extracting All the Goodness in a Tea

In order to extract all the yummy medicinal properties from our turkey tail mushrooms, we need to simmer them for a long time. Traditional recipes for making turkey tail tea call for at least 90 minutes of simmering, and some for much more. I usually aim for 3 hours.

The average daily dosage of mushrooms is 0.3 to 0.4 ounce (9–11 g), so to make enough tea for a whole week I simmered around 1.2 ounces (34 g) of mushrooms in 6 cups (1.4 L) of water. Due to evaporation, after 3 hours I ended up with around 4 cups (1 L) of tea.

TURKEY TAIL MUSHROOM SODA

Turkey tail (*Trametes versicolor*) is a common polypore mushroom found throughout the world. The common name of this mushroom comes from the fact that it resembles an actual turkey tail. It's found on dead and fallen trees or stumps. It's an extremely medicinal mushroom used presently in holistic or naturopathic medicine for its anti-cancer properties and ability to boost the immune system.

It's usually made into a tea, but you can also use this mushroom to make medicinal sodas or beers. The tea itself isn't fantastic, so I like to add other ingredients such as lemons and ginger when I make a soda, but there are tons of creative combinations you can try. I also use these mushrooms quite a bit as an added (and medicinal) bittering agent in some of my primitive beers. If you play with them for medicinal uses, make sure to do some research on daily dosage.

4 cups (1 L) turkey tail tea
4 cups (1 L) water, to add to the tea
3 lemons, sliced
0.5–0.7 ounce (14–20 g) finely cut gingerroot
3.5–4 ounces (100–125 g) sugar, or honey
⅓ cup (80 ml) ginger bug or other wild yeast starter (or use wine/champagne yeast)

Procedure

1. First, pour the tea into a container, then add the water, sliced lemons, and ginger.
2. Next, add the sugar and shake the jar a few times to make sure the sugar is diluted.
3. Add ⅓ cup (80 ml) of ginger bug or a similar wild yeast starter. You can also use champagne yeast (follow the manufacturer's instructions for around ½ gallon or 2 liters).
4. Close the container but not too tightly; you want fermentation gases to escape. Three times a day, shake the liquid for a few seconds. Within a day or two you should have a nice fermentation going.
5. When the fermentation is going well, transfer the liquid into a recycled soda bottle or a glass swing-top bottle and let it ferment for 8 hours or so. Check the pressure by opening the top carefully and slowly. If you're happy, place the bottle in the fridge. The cool temperature will slow down the fermentation process considerably. Drink within a week; about one-fifth of the liquid each day would be the proper dosage.

Turkey Tail Mushroom Soda brewing.
Turkey tails are known to boost the immune system
and are used in naturopathic medicine
to reduce cancer tumors.

CHAGA BEER

Chaga is a parasitic mushroom found predominantly on birch trees. Similar to turkey tails, it is used to support the immune system and has been used as a folk remedy to help treat various conditions such as high blood pressure, high cholesterol, stomach diseases and ulcers, diabetes, and others. I use it as a medicinal bittering agent in some of my beers along with other medicinal mushrooms such as turkey tail or reishi.

This beer recipe uses the same average amount of mushroom as a traditional chaga tea. You can use the same procedure to make turkey tail or reishi beer. I use around 2.4 ounces (70 g) of turkey tail and 2 ounces (60 g) reishi. One gallon should give you a bit more than seven 16-ounce (500 ml) swing-top bottles. You can drink one bottle of beer per day.

1 gallon (3.78 L) water
0.3 ounce (9 g) mugwort, or 0.15 ounce (4 g) dry hop flowers (I used Cascade hops with 5 percent IBU)
0.07 ounce (2 g) willow bark (optional)
0.07 ounce (2 g) yarrow flowers (optional—for flavor)
1¼ pounds (567 g) organic brown sugar
3 lemons, or 1 cup (80 g) lemonade/sumac berries
2.1 ounces (60 g) chaga mushroom
Commercial beer yeast, or ½–¾ cup (120–180 ml) wild yeast

I usually make this drink in two stages. First I make a traditional chaga tea and strain the liquid. I add this to my beer later on, after the boiling process.

Making the Chaga Tea

If my chaga comes in large chunks, I like to break it down into smaller bits. I use the stone (*mano*) from my Mexican stone grinder (*metate*) for this, but you could place the large bits into a towel and use a hammer to break them down. I like smaller bits that are around ½ inch (1.3 cm) in diameter or so. Some people go further and reduce the small bits into powder, but I'm not sure that's necessary.

Place the chaga pieces into a pot with around ½ gallon (1.89 L) of water and simmer the solution for 3 to 4 hours. Once this is done, remove the pot from the heat and strain. You'll need to measure how much tea you have. I usually end up with 5 to 6 cups (1.2–1.4 L) of very dark liquid. You deduct that amount from a gallon (3.78 L) so you know how much water you'll need to make the beer. For example, if I end up with 6 cups (1.4 L) of tea, I would use 10 cups (2.36 L) of water for the next step.

Breaking down the chaga chunks into smaller bits and making a coarse flour using my stone grinder (metate).

Making the Beer

At this point it's pretty much the same procedure as making a regular (wild) beer—but note that you add the chaga tea after the other ingredients have been boiled. (See "Hot Brewing" on page 81.)

1. Mix the water, mugwort, willow bark, yarrow flowers, and sugar in a large pot. Cut and squeeze the lemons into the solution. Place the lid on and bring the mixture to a boil. Let it slowly boil for 30 minutes.
2. Remove the pot from the heat and add the chaga tea. Cover the pot again.
3. Place the pot in a pan of cold water; cool to 70°F (21°C), then add the yeast (wild or commercial). When I'm using a wild yeast starter, I usually use ½ to ¾ cup (120–180 ml) of liquid.
4. Strain the brew into the fermenter. Position the airlock or cover your fermenter with a paper towel or cheesecloth. Let the beer ferment for 10 days. Start counting when the fermentation is active (this may take 2 to 3 days with a wild yeast starter).
5. Siphon into beer bottles and prime each bottle with ½ teaspoon (2 g) brown sugar for carbonation. Close the bottles and store in a place that's not too hot. The beer will be ready to drink in 3 to 4 weeks.

CHAPTER 9

Naturally Fermented Sodas

This was the ingredients list of a random soda bottle I picked up at my local supermarket yesterday, and pretty much every other label was the same. The main differences among the various brands were the artificial colors and flavors.

Carbonated water
Artificial colors
High-fructose corn syrup
Caffeine
Citric acid
Sodium benzoate
Potassium sorbate

It's amazing to see just how far many modern commercial sodas have moved from their healthful lineage. Like root beer, many sodas were originally created by pharmacists and marketed locally as healthy or medicinal beverages. Granted, you had exceptions in terms of "healthy drinks"—some recipes included ingredients such as cocaine or lithium citrate—but otherwise many early sodas were created for the sake of body health. That heritage lingers in some popular commercial soda names such as Dr Pepper.

As sodas became more popular, the quest to make cheap, tasty, and highly profitable drinks started, and in the process turned potentially healthier probiotic beverages into the artificially flavored sugary monstrosities we find today in our supermarkets. As a historical note, not all early sodas were fermented products; quite a few used an acid such as citric acid mixed with bicarbonate of soda to create carbon dioxide gas. This generated bubbles of carbon dioxide, which made the drink fizzy.

But from a healthier fermentation perspective, it's quite sad that most people have forgotten how simple and natural sodas were made: some wild yeast, a natural sugar source like honey or maple syrup, and some flavorful ingredients are all you need to create a fizzy, tasty drink. Today, and more than ever, making your own sodas at home is a much healthier alternative for yourself or your family than the sodas you buy at local stores.

The old way of making soda is really a very simple fermentation process. You make juice or an infusion, add a sugar source and some wild or commercial yeast to it, let it ferment for a short time in a clean container (carboy, large bottle, jar, or the like), then transfer it to a closed bottle (plastic soda or swing-top glass bottle). With experience, you can even simply pour your sugary liquid directly into a swing-top or recycled soda bottle, add yeast, and close the bottle.

It's a probiotic process with live yeast. As fermentation continues within the closed bottle, pressure is building, and this is what creates the natural carbonation.

You can start with extremely simple ingredients such as a mint or fig leaf syrup (see the "Flavored Syrups" section that begins on page 21).

For example, this morning I made some soda by simply placing ½ cup (120 ml) of mint/lemon syrup inside a 1-quart (1 L) recycled soda bottle, filled it with water, and added 2 tablespoons (30 ml) of ginger bug starter (wild yeast). By tomorrow I should have a nicely carbonated tasty drink. It's that simple!

You can approach making sodas from various perspectives. If you're a chef, they make an excellent nonalcoholic alternative for your guests. When I was working with Chef Ludo Lefebvre, I probably designed more than 15 different natural soda recipes. If you're a bartender, you may be interested in fermenting bitter sodas as an addition to cocktails. If you're an herbalist, your interest would be in fermenting healing herbs, which is pretty much how sodas got started in the first place.

One tip: Don't use beer or bread yeast to ferment your soda unless you want it to have some beer or bread flavors. I usually use wild yeast or champagne yeast. Wine yeast works well, too.

THE BASIC SODA PROCEDURE

I make all kinds of sodas and, although the basic principle is the same, there are slight variations in the preparations. The ingredients you use will dictate the best method. Some plants, such as mint, basil, pine needles, or (locally) yerba santa (*Eriodictyon californicum*), are much better in a cold infusion, while others—dehydrated berries, various sages, mugwort—are better boiled first to extract the flavors.

Feel free to experiment with plants from your garden or local terroir using both methods—it's all about flavor.

You don't need a lot of equipment to make soda. Here is the basic list for making a 1-gallon (3.78 L) batch using the hot (boiling first) method detailed on page 243.

2–3 cups (400–600 g) sugar, or 2–3 cups (500–700 ml) honey—you can also use less

Yeast (champagne or wild yeast)

1 gallon (3.78 L) water—don't use tap water, which may contain chlorine

Large pot with lid (if you boil the juice/infusion)

1-gallon (3.78 L) bottle or glass container

Sieve

Funnel

Soda bottles (recycled plastic bottles or swing-top glass bottles)

Measuring cup

Airlock and stopper (available online or at local brewing supply stores; you can also top the gallon bottle with a clean paper towel and a rubber band)

Procedure

Make your sugary tea/brew, infusion, or juice, then place it into a bottle or container with an airlock or covered with a clean towel. Let it ferment for around 24 hours. (If you've used a wild yeast starter, this may take a couple of days.) The airlock is there to let the CO_2 gas escape (a by-product of the fermentation process) and to prevent anything from getting inside (fruit flies, bacteria, or the like).

After around 24 hours of fermentation, strain and transfer your soda to closed bottles and wait for another 8 to 24 hours. Pressure builds up inside the bottles and creates the carbonation. At that point you can place your soda in the fridge, where the low temperature will slow the fermentation process.

There is a bit of an art involved in getting the right amount of carbonation in the bottles. If the original fermentation is very active (lots of bubbling), you don't need to leave the soda in the closed bottle for 24 hours—that may be too long, and you'll get lots of pressure. This isn't always a good thing; you may open the bottle and experience a gusher.

I use the airlock to see how active the fermentation is. If I see a bubble going through every second, I leave the soda in the closed bottle for maybe 8 hours, then place it in the fridge. If I see a bubble going through every 2 seconds, I usually leave the soda in the bottle for 16 hours. You can always check the pressure by opening a bottle carefully and very slowly. See "Wild Carbonation in Bottles" page 99, for more tips on pressure and carbonation.

Note that as you gain experience, you may be able to skip a lot of equipment and steps. For example, I often make super-quick sodas for my classes and workshops by placing ½ cup/120 ml of honey, flavored syrup, or cane sugar in a recycled soda bottle, pouring in some water, putting some mint leaves or various other tasty plants inside with a bit of lemon juice, and adding 2 tablespoons (30 ml) of yeast starter. By the next day my bottles are carbonated. I do a quick check for pressure, then place them in the fridge. From experience, it's much better if you leave your soda in the fridge for at least a few hours or overnight before enjoying it. Something seems to make the carbonation "settle into" the liquid this way.

Various sodas in progress—bottled and fermenting.

MAKING SODA: THE BOILING METHOD

Making a soda using the hot method is very similar to making sugar-based beers. You boil the ingredients, cool the solution, add your yeast, let the mixture ferment for 18 to 24 hours, then bottle it.

As you experiment with foraged or purchased ingredients, you'll discover that boiling your solution works in some cases, but can also alter the taste too much with certain specific herbs. For example, you lose a lot of subtle flavors when cooking mints, chervil, fennel, basil, or white fir, whereas herbs like mugwort, yarrow, or dehydrated berries work very well. You don't always have to boil the ingredients, either; you can make some infusions by boiling the water, removing it from the heat, then placing your herbs in the hot water and letting them steep for a specific period of time. Once the solution has cooled down, you add your yeast, then strain and place in your fermenting vessel.

Here's a basic recipe for a ½-gallon (1.89 L) batch of wild berry soda.

½ gallon (1.89 L) water—don't use tap water, which may contain chlorine

2 lemons

½–¾ cup (65–90 g) dehydrated blueberries (feel free to try other berries)

½ teaspoon (1 g) wild fennel seeds (or buy commercial fennel seeds at the store)

Small handful (1–2 g) dried mugwort

½–¾ cup (100–150 g) organic white sugar

¼ cup (60 ml) honey

Champagne yeast (or wild yeast starter)

Procedure

1. Pour the water into a pot and add all of the other ingredients except the yeast.
2. Bring the solution to a boil, then simmer for 20 to 30 minutes.
3. Remove the pot from the heat and place it in a sink filled with cold water. You may need to change the cold water a few times; you want the solution to be lukewarm (70°F/21°C). Add the yeast. (If the liquid is too hot, it will kill the yeast.)
4. Clean your fermentation bottle thoroughly. You can find ½-gallon or gallon bottles at most regular grocery stores, usually cheap wine bottles. Place a sieve and funnel in the bottle's neck, then pour in the solution. With very clean hands, you can also squeeze the ingredients left in the sieve/funnel to extract more flavors.
5. Open your champagne yeast packet (a 5-gram packet is usually good for making 5 gallons/18.9 L) and pour some of the yeast inside the bottle— enough for ½ gallon, which isn't much.

Gather all the ingredients you'll need to make your soda—dried berries, herbs, fruits, and so on.

Boil your ingredients to extract the flavors—usually 20 to 30 minutes.

Remove the pot from the heat and cool the liquid.

Strain your soda into your fermentation vessel, usually a bottle.

6. Clean the stopper (cork) and airlock. I usually use a quick rinse in very hot water. From time to time I'll sanitize my airlocks in a diluted bleach solution (1 teaspoon/5 ml bleach per 1 gallon/3.78 L water for a few minutes) if they get dirty.

   Set the stopper and airlock in place, and wait 24 hours. Usually after 10 to 12 hours, sometimes sooner, you will see fermentation activity and gas escaping in the form of bubbles. That's what you want.

7. Pour your fermenting soda into recycled plastic soda bottles or swing-top glass bottles. Close the bottle tops. If your fermentation was very active, check the pressure after 8 hours by opening the top slowly and carefully. If there isn't enough pressure, keep fermenting the soda and check again after another 8 or more hours. When you're satisfied, place the bottles in the refrigerator, which greatly slows the fermentation process.

Add the yeast—commercial yeast or wild yeast starter. For sodas, you can use champagne or wine yeast.

Place the airlock and ferment for around 24 hours, then bottle.

COLD-INFUSED WILD SODAS
(HIKE SODAS)

This method is fantastic for making quick sodas out of your favorite hiking place or to study the flavors of whole environments.

Unlike most beers, you really don't need to boil most of your ingredients with "wild" sodas. The fermentation is rather quick—I usually ferment my cold-infused sodas for 24 to 48 hours at the most. During that time, the yeast that you added to the brew will take over and, for now, be the obvious winner in this war against unwanted bacteria. Over time, this could possibly change, but because you'll drink the soda pretty soon after the initial fermentation, the potential spoilage won't happen. Enjoy the taste of victory!

As I explained in "Making Soda: The Boiling Method" (page 243), I choose whether or not to boil my ingredients based largely on flavor. Some herbs, such as mugwort and yarrow, unleash their flavors through boiling, but other herbs or plants including pines, fir, grass, fennel, and mints are much better when enjoyed fresh. If you boil mint, the flavor changes considerably. I really like the taste of some fresh pine or white fir needles, but when boiled . . . not so much.

It's clearly a question of individual preferences, though. As you experiment with wild brews and sodas, you'll figure out the appropriate methods. That's part of the fun.

Aside from flavors, some sodas are made for medicinal purposes. A pine needle and lemon soda, for instance, is packed with vitamin C. Often vitamins and nutrients are altered through the boiling process, however, making a cold infusion more appropriate.

Procedure

1. Decide on the ingredients you'll use to create your soda—herbs, fruits, aromatics, juices, and anything else—and forage or purchase them.
2. Clean them thoroughly, as well any container and utensils you will be using. You can go to the extreme of placing your container in boiling water to pasteurize it, but I've never had any problem with just a thorough cleaning with hot water and regular dish soap.

3. With clean hands, place your ingredients into the container. Add water (not tap water, which may contain chlorine) and a sugar source (white sugar, honey, brown sugar, molasses, birch or maple syrup, and so on). Mix everything so your sugar base is well diluted and then add the yeast. You can use either a yeast starter (½–¾ cup per gallon, or 120–180 ml per 3.78 L) that you have made yourself from wild berries, or a commercial yeast strain. I often use

Hike in the Forest Soda: organic lemons, wild mint (similar to spearmint), a tad of yarrow, white fir, native cherries, and white sugar.

Spring Forest Study: wood sorrel, forest grass, mugwort, cranberries, blueberries, willow leaves, organic lemons, sagebrush, and black sage, with white sugar as a sugar source.

champagne yeast purchased online or at the local brewing supply store.

For 1 gallon (3.78 L), I use between 1 and 1½ cups (200–300 g) sugar or honey (236–354 ml). You can experiment with less, or add more if you want a very sweet soda. Most soda recipes call for 1½ to 2 cups (300–400 g) of sugar per gallon, so I tend to be on the low-sugar side.

4. Place a paper towel on top and secure it. If you used a jar, simply screw the metal band over the paper towel. Stir three or four times daily with a clean wooden spoon or other instrument.

5. Wait around 24 hours (or up to a couple of days if you used a wild yeast starter); at this point the fermentation bubbling should be quite obvious. Taste! If you like it, you can strain the fermenting solution into recycled plastic soda bottles or swing-top bottles, then wait a day and check the pressure. If it's appropriate, you can put your soda in the fridge.

Hike in the Chaparral Soda: yerba santa, black sage, manzanita berries, wild fennel, woolly bluecurls flowers, organic lemons, water mint, California sagebrush, and juniper berries. Local honey and white sugar.

Trip to Gloria's Place in the Mountains—Angeles Crest Creamery: pinyon pine, white fir, California juniper, manzanita and lemonade berries. Pinyon pine syrup and mountain honey are the sugar sources.

MAKING SODAS WITH FRUIT JUICES

If you don't have time to go foraging or pick up ingredients at the local farmer's market, you can make excellent sodas using fruit juice purchased at the local store. Sometimes I make sodas using organic fruit juice.

Be aware that some (usually cheap) juices aren't really juices but a mix of high-fructose corn syrup, colorants, preservatives, chemicals to preserve flavors, juice from concentrate, et cetera, and they may not ferment for obvious reasons. Even yeast wouldn't touch them!

I usually don't boil the juice (most of them are already pasteurized) unless I add other ingredients. For example, if I make a blend of blueberry juice with a bit of mugwort, I would take a portion of the juice—maybe 20 percent—boil a few dry mugwort leaves in it, then let it cool before straining it and pouring it back into the original juice.

To be honest, I've also made sodas with freshly made juice using the cold infusion method. Fermentation is a very safe process. If a fermentation goes bad, the resulting liquid will not smell good—it's definitely not something you'd be tempted to use. In my 10 years of fermenting, though, I've never had one go bad. Just use good hygiene, wash your hands often, clean the containers and bottles you're using, and you should be fine.

Some purchased fruit juices already have sugar in them, so you may not need to add any more. I do it by taste. If a fruit juice is very sweet, I won't add sugar. If it's not too sweet, I may add ¼ cup (57 g) of white sugar or ½ cup (120 ml) of honey per gallon. Use your judgment.

Procedure

1. Clean your fermentation bottle and pour in the juice. (If the juice is very sugary, I often I add water—around one-fourth the volume.)
2. Open your champagne or wine yeast packet (good for fermenting 5 gallons) and pour some of the yeast into the bottle. You can also use a wild yeast starter (½–¾ cup per gallon, or 120–180 ml per 3.78 L).
3. Clean the stopper (cork) and airlock. I usually do a quick hot-water rinse. From time to time, I'll clean my airlocks in a bleach solution if they get dirty (1 teaspoon per gallon for a few minutes; rinse thoroughly afterward).
4. Set the stopper and airlock in place, and wait around 24 hours. Usually after 10 to 12 hours, sometimes sooner, you will see fermentation activity and gas escaping. That's what you want. By the way, if your fermentation is really active before 24 hours, it's completely okay to proceed to the next step; it's not unusual for me to bottle some of my fermenting juice after 16 hours or so. If you used a wild yeast starter, on the other hand, it may take a bit longer for the wild yeast to take off, often 2 to 3 days depending on the temperature.
5. Pour the fermenting juice into plastic soda bottles or swing-top glass bottles. Close the bottle tops. Check the pressure from time to time.
6. When you're satisfied with the level of carbonation, place your soda in the fridge and enjoy later on.

If you're using purchased juices, you can also simply ferment them directly in recycled soda bottles. Often I just transfer my juice into a clean bottle, add a bit of yeast, and check the pressure after a day or so. If I'm happy (the pressure should feel like a regular bottle of soda) then I set the soda in the fridge to slow the fermentation process.

Rural Wilderness Drink: Organic pomegranate and blueberry juice, lemons, turkey tail mushrooms, woolly bluecurls flowers, sugar bush berries, and mugwort.

Regular Fruit Juice and Wild Ingredient Sodas

Even if you purchase fruit juices from your local market or buy fresh fruits (pomegranates, apples, watermelons, oranges) and make the juice yourself, you can still add some local flavors to your drink by adding foraged ingredients from your garden or wilderness.

By the way, don't always think of the wilderness as something remote from the city. Although I'm in Los Angeles, I actually live in a semi-rural environment, and a simple walk through my neighborhood allows me to forage the ingredients you see in the photos. Just be polite and ask permission if you're picking from private land. Bring the owners some soda (or beer) later on and you'll make a lot of friends. It's a good way to get access to large private properties.

I often use wild berries to add interesting flavors. For example, recently I made a fruit juice soda (cherry) and added dehydrated blueberries, cranberries, a few sage leaves, and a couple of lemons. If you're using garden plants, adding a bit of rosemary or thyme to blueberry juice is quite fantastic.

My new favorite soda is made using a locally purchased pomegranate/blueberry juice and plants found in my garden and neighborhood. The juice is already pasteurized. I simply pour most of it into a clean 1-gallon (3.78 L) bottle and set aside around 2 cups (0.5 L). I don't need to add sugar to the juice, as there is already enough.

I pour the 2 cups (0.5 L) of juice into a pot, then add my local plants (woolly bluecurls, mugwort, turkey tail mushrooms, sugar bush berries) to the pot as well.

I bring the liquid to a boil, then simmer for around 10 minutes. The pot is then placed in cold water with the lid on. I change the cold water two or three times until the juice is lukewarm (70°F/21°C).

My next step is to strain the juice into the bottle, using my (clean) hands to squeeze the ingredients so I get all the flavors in there.

The final step is to add the yeast (commercial yeast or wild yeast starter) and place the airlock on top of the bottle.

After a short fermentation, between 12 and 24 hours depending on how active it is, I bottle the soda and check the pressure after 8 hours or so. If I'm satisfied, I place it in the fridge. If I've used a wild yeast starter, it may take 2 or 3 days for the wild yeast to take off.

When you mix regular juices and wild ingredients, the number of interesting flavors you can create is infinite.

Fermenting Directly in Your Final Bottle

I consider this an advanced fermentation technique, because you skip the step of placing your juice/infusion in a fermenter (jar or large bottle) before bottling it, instead fermenting your soda in the final bottle (recycled soda bottle or swing-top bottle).

This method works well if you don't have a lot of solid ingredients (such as barks and leaves). These days I use it all the time—*but* I also have years of experience behind me and can judge how active a fermentation will be based on the local temperature, the amount of sugar I used, the ingredients I chose, and the mood of local fermentation gods that day.

The danger in skipping the intermediate step of fermenting in another container before bottling the liquid is the possibility that you can fail to see how active your fermentation really is. You might end up with a bottle explosion (if you used a glass bottle) or a gushing soda when you open the bottle.

By fermenting in a large bottle or container before you put your fermentation into its final bottle, you're able to judge how active it is by watching the various signs: bubbling, movements within the liquid, and, most important, the speed of gas emission through your airlock. That's why I like to use airlocks; it's a great way to gauge fermentation activity.

With time and experience and by watching the airlock, you'll even be able to judge how long it will take to get a good carbonation going once your brew is bottled, and when to place the bottle in the fridge without having to open it even a tiny bit to check the pressure.

Do this for a few months, and fermentation becomes part of who you are . . . you just know. These days, I make a lot of simple sodas by fermenting them in the final bottle. It's easier and quicker. If you do the same, place the soda in the fridge before serving it, at least for a few hours and preferably overnight or longer. I've found that it helps "settle" the carbonation, which means you can avoid spraying liquids on your guests.

How Much Sugar Should You Use to Make Sodas?

I usually use around 1½ to 3 cups (300–600 g) of sugar or 1½ to 3 cups (360–700 ml) of honey per gallon (3.78 L). Again, there are no real rules; you can experiment with less, but you will need some sugar if you want to get fermentation going, and if you don't use enough, your soda can taste a bit sour because the yeast will have consumed most of it (though you can always add some during the process or at the end). Feel free to increase the sugar if you like very sweet sodas.

UNRIPE PINECONE SODA

This recipe really came from a happy accident. As I explained in the beginning of this book, while I was once working on a recipe for unripe pinyon pinecone syrup, I discovered that the cones were loaded with wild yeast—so loaded that my jar exploded within a day.

Ever since that magical yeasty discovery, when springtime comes these unripe cones are one of my favorite ways to make yeast starters. All you need is springwater, some sugar, and unripe pinyon pinecones. (My other favorite yeast starter is based on California juniper berries. The recipe is probably the simplest one in the book, and is a bit similar to making the fermented juniper berry drink smreka on page 214.)

But if you don't have green (unripe) pinyon pinecones, can you use other cones?

Well, a while ago I posted about my wild yeast starter experiments on various online fermentation groups. I got quite a lot of feedback from people who had experimented with their own local unripe pinecones with success. I would say that 80 percent of them managed to do it successfully.

All true pines (*Pinus* spp.) are technically edible, so you're pretty much free to experiment with your local unripe pinecones. If you want to make soda,

though, you'll need to find some flavorful ones (white pine is good). Not all pines have much flavor.

The idea of making soda with unripe cones came from my wild yeast starters. These smelled so good that from time to time I would actually drink some, and one day I decided to try making soda from one directly. With unripe pinecones, you judge by flavor, so don't hesitate to taste during the fermentation process and stop when you like your brew. Ferment too long and the pine flavors can be overwhelming—though even then it can be used as an excellent blend for making cocktails. I'm sure there are a lot of possibilities I haven't explored yet.

Procedure

1. Place ¾ cup (150 g) sugar in a ½-gallon (1.89 L) jar, fill it up with springwater, and drop in a couple of unripe pinecones.

2. Screw on the lid, but not too tightly; you want fermentation gases to escape. Three times a day, shake the jar for 10 seconds or so. I usually get a fermentation going in 2 or 3 days.

3. Leave the cones inside the jar and continue the ritual of shaking three times a day. When the fermentation is going well, start tasting. It's very much like making lazy wine (page 194): Work with it as you go along, adding sugar if you want more alcohol. Stop when you like what you have, then strain and pour the liquid into recycled plastic soda bottles or glass swing-top bottles. Check the pressure after a day or so (see "Wild Carbonation in Bottles" on page 99).

   When you're satisfied with the level of carbonation, place your brew in the fridge and enjoy it the next day. I like to drink it within a week.

Pinyon Pinecone Soda served with water mint and local wild cherries.

MOUNTAIN
RASPBERRY/BLUEBERRY SODA

This is probably the prettiest soda I've ever made, and it's super tasty as well. This soda is less about a specific location than it is about the season.

Here in Southern California, fall is a time when everything turns into a desert, which means that as a forager I usually have to go to higher altitudes where the temperature is lower to look for fresh flavors. From my perspective, the local mountains in late fall are all about pines, fir, and spruce, but if you visit your local grocery store, this is also a time when berries are extremely abundant. With this recipe, I wanted to celebrate what nature and local organic farmers can offer in autumn, and it was a great excuse for spending a whole day outdoors.

My very first stop was at the local farmer's market, where I picked up the nicest and freshest raspberries and blueberries. I wish I could forage blueberries in Southern California, but my best option is to support my local small organic farmers. For them, it's really a labor of love, and the berries available in regular grocery stores can't compare in flavor. My berries were super juicy and sugary.

Next was a trip through the local mountains to visit my friend Gloria and say hi to her goats and a new donkey. There I collected some pinyon pine and white fir branches. With pinyon pine, I'm extremely interested in the tasty sap, so I like to use actual small branches cut into small sections (1 inch/2.5 cm or so) to infuse flavors.

The recipe is extremely loose, so don't overcomplicate things. If you live in Oregon or on the East Coast (Vermont, Maine, New York), you may use different trees such as spruces or pines (white pine, blue spruce, and so on), and of course different berries. If you live in Maine or Vermont, by all means use your wonderful maple syrup as a sugar source.

It's all good. This is very much a "concept" drink to celebrate the season.

Procedure

1. Clean your fermentation vessel and place your fresh ingredients in it. Add springwater (don't use tap water, which may contain chlorine) and sugar. There are no real rules for this simple fermentation: Just pack 80 percent of your jar with what you have collected/purchased. I use around 50 percent berries and 50 percent pine/fir. Cut the tips of the needles to help extract the flavors. Start with around ½ cup (100 g) of sugar for ½ gallon (1.89 L) of water.

2. Add yeast (wild yeast starter or champagne yeast). Screw on the lid of your container, but not too tight; you want fermentation gases to escape. You can also cover it with clean cheesecloth or a paper towel. Stir gently two or three times a day with a clean spoon.

   Taste as you go, and judge by flavors. When you like what you drink, you can stop. It may take 3 or 4 days or more. I usually don't leave the fruits in the liquid for more than 3 days (some get mushy).

3. Strain and pour the liquid into recycled plastic soda bottles or glass swing-top bottles. Check the pressure after a day or so (see "Wild Carbonation in Bottles" on page 99).

4. When you're satisfied with the level of carbonation, place your soda in the fridge and enjoy it the next day. I like to drink it within a week.

PINE NEEDLE SODA

If you've ever tasted some delicious pine needles, trust me, you'll want to brew this soda. My favorite pines are pinyon pine, ponderosa pine, and white pine, but I've also made blends that included white fir and spruce. In fact, when I was teaching in Vermont, we made a similar soda using white pine needles and blue spruce tips. Note that ponderosa pine and white fir are not recommended for consumption if you're pregnant.

My regular mix is usually composed of mostly pinyon pine needles (60 to 80 percent of the blend) and some white fir needles (20 percent) with one or two lemons. I make sure to cut the pine and fir needles with scissors so they can release their flavors quickly. I slice the lemons into five or six parts but, if you're an experienced forager, you can use sumac or lemonade berries instead.

Procedure

1. The method is similar that for the Elderflower–Pineapple Weed Soda (page 262). Fill around half of your (clean) container loosely with the ingredients, add some springwater and 1 to 1½ cups (225–335 g) organic cane sugar or honey, then add the yeast and place a paper towel on top secured by a rubber band or a string.

2. Using a clean wooden spoon, stir the liquid three or four times a day until you get a nice fermentation going—this usually takes 2 to 3 days in Southern California.

3. Strain the liquid into recycled soda bottles and check the pressure after a day or so, then refrigerate for at least 8 hours before enjoying. With pine sodas, you can really judge by flavors; taste as you go along and stop the fermentation whenever you're satisfied.

ELDERFLOWER–PINEAPPLE
WEED SODA

This soda is simple to put together and a good representation of my local terroir during late spring.

During that time of the year, two of the most abundant and flavorful wild edibles are pineapple weed and elderflowers. Pineapple weed is a type of wild chamomile found across North America; the flowers actually look and taste like pineapple. They're wonderful for making infusions or fermented concoctions. I even use them in some of my primitive beers.

You really can't go wrong with the flavor combination of elderflower, lemon, and pineapple. It makes a delicious fruity and floral soda.

2 cups (500 ml) pineapple weed flowers

4–6 cups (1–1½ L) elderflowers

3 lemons

1 gallon (3.78 L) springwater

1½ cups (335 g) organic cane sugar or honey

¾ cup (180 ml) wild yeast starter (or champagne yeast)

Notes: *Elderflowers contain a lot of small bugs. I usually leave my elderflowers outside in a bowl for a few hours; most of the bugs will vacate during that time. But the reality is, you'll need to filter that soda when you bottle it unless you like the extra bug protein.*

Do not place your elderflowers in water beforehand to "clean" them. A lot of the flavors and wild yeast are in the pollen.

As with many other soda recipes, you can make a boozier drink by using more sugar and fermenting longer.

Procedure

1. I fill half of my (clean) container loosely with the flowers and lemons, add the springwater and sugar, then add the yeast and set a paper towel on top secured by a rubber band.

   For this soda, I use a wild yeast starter, but it's not always necessary—the elderflowers have wild yeast as well.

2. Using a clean wooden spoon, I stir the liquid three or four times a day until I get a nice fermentation—usually 2 to 3 days in Southern California. I don't leave my ingredients in the fermenting water more than 3 days, as I think this adversely affects the flavors.

3. My next step is to strain the liquid (see the notes) into another container and ferment it a bit longer. But you can also stop the fermentation right away if you're happy with the flavors.

   Regardless, after fermentation pour the liquid into recycled soda bottles and check the pressure after a day or so. Put the soda in the fridge for at least 8 hours before serving. I usually drink it within a week or so.

BRANCH AND TWIG SODA

During winter and early spring, the wonderful flavors I usually find in white fir, pinyon pine, and white and ponderosa pine needles are very subdued; more often than not, in fact, the needles become too bitter for any culinary uses. It's as if the trees become dormant and the life force (and flavor) retreats deeper, maybe hibernating, as a protection against the elements.

My approach used to be to wait until summer before I would use the needles in my cold infusions, sodas, and beers. This changed one year when I realized that, even during the colder months, a lot of the flavors were still available within the branches. It's especially true with pinyon pine—around February the needles are still quite bitter, but the branches (and wood inside) are loaded with aromatic and tasty sap.

You really don't need to cut a lot of branches to make a drink, and in fact you're not hurting the tree in any way by cutting off just the end part of a branch. According to forestry experts, you can even remove the lower third of a crown without harming a pine tree. Based on their advice, cutting is best done in late winter before the growth phase begins, and it's precisely the best time to do it for flavors, too. Make a clean cut; don't break off the branch with your hands.

The method I use is very simple. Using my pruning shears, I cut the end of a branch where it's around ¾ inch (2 cm) thick and can provide me with a segment around 7 to 10 inches (17–25 cm) long. When that's done, I also remove most of the smaller side branches where the needles are located.

Once I'm back home, using my pruning shears again I'll cut the branches into smaller segments to fit my fermenting container, usually 3 to 4 inches (7.5–10 cm) in length. Then I make an incision at one end and, using my fingers, pry the branch/twig open so I end up with two separate parts and the wood inside (and the flavors) exposed. The smell is intoxicating, with hints of tangerine, pine, lemon, and orange blossoms.

As you can see in the photo, you really don't need a lot of branch tips, and usually 24 to 48 hours is enough to extract their flavors. Do it longer and it can be too much. I taste as I go along and stop when I like it.

½ gallon (use 2 L) water
½–¾ cup (100–150 g) white sugar, or honey
7–8 cracked California juniper berries (unripe)
3 small pinyon pine branches
    (3–4 inches/7.5–10 cm long), separated
    into two segments lengthwise
1 small branch white fir with the top of
    the needles cut
½ lemon, sliced (optional)
1 dried lime (see page 168), cracked open
½ cup (120 ml) wild yeast starter
    (or champagne yeast)

Procedure

1. Place everything in a clean container, then add the wild yeast starter. Shake the jar for a few seconds to make sure the sugar is dissolved, then screw the lid on (but not too tightly; you want fermentation gases to escape). I shake the container every 6 hours or so.

2. Around 18 hours of infusion is usually enough. The fermentation has barely started, and you can't see any bubbling going on, but strain the liquid into a recycled soda bottle and let the fermentation continue inside the bottle, which will create the carbonation. I monitor it by pressing the side of my soda bottle to see how much pressure has built up inside. You can also open the bottle slightly if you want. Once you're satisfied with the carbonation, place the bottle in the fridge, which will slow the fermentation process. I usually use the soda within a week, but it will last much longer. Just remember to check the pressure from time to time.

SoBeer

I've extracted this recipe from my book *The New Wildcrafted Cuisine*. The result of this kind of fermentation tastes like a cross between a soda and a beer. Some people could even call it a type of "small beer" (one with a very low alcohol content).

The idea came to me out of necessity. We had an unplanned private event and the client wanted to have some wild brew on the menu. Usually my beers need a minimum of 3 to 4 weeks of aging in the bottles before they're ready for drinking, but in this case I only had 3 weeks from preparation to actual consumption.

I've experimented a lot with sodas and native brews, which don't require a lengthy fermentation, so I decided to try a new method to meet the deadline. It's probably the same way some low-alcohol beers are made.

The recipe for my most popular beer (mugwort and lemons) calls for 1¼ pounds (567 g) of brown sugar for 1 gallon (3.78 L). The initial fermentation in the carboy (the large bottle used for ferments) is 10 days. Based on the fact that 1 pound (454 g) of sugar would give me a rough estimate of 5 percent alcohol, I concluded that I could make a lower-alcohol beer by cutting down the brown sugar to 10 ounces (284 g); I'd end up with a beer of around 3 percent alcohol and the regular flavor. The fermentation would take less time because there was less sugar.

It was the only choice I could come up with. So I made a gallon of beer with half the sugar, fermented it for 6 days in the carboy, and bottled it for 2 weeks. It worked! People loved it.

Since then, I've gone further and made all kinds of low-alcohol wild brews with even less sugar and fermentation time. When a reporter from *LA Weekly* magazine wanted to interview me for an article about the wild brews I created, I managed to make a tasty forest beer composed of seven wild ingredients within a week by using 7 ounces (198 g) of sugar, fermenting it for 5 days, then bottling it for only 2 more days. The end result was very similar to a sour, fruity Belgian beer with some hints of grapefruit.

The reporter asked me what it kind of beer it was and, not knowing how to properly name the thing, out of the blue I told her it was a SoBeer: a low-alcohol primitive brew that's a cross between a soda and a beer. I actually like the name as it doesn't sound too serious, so I've been using it ever since.

The more I get into primitive fermented brews, the more I realize that, aside from the basic preparation methods, there are no real rules—it's all about flavors. If you can make a wild brew that tastes like a low-alcohol beer in 10 days, why not? A lot of native brews were simply fermented for a few days with wild yeasts, then enjoyed.

The SoBeer in the photo has the following ingredients for 1 gallon (3.78 L): a small handful of dry mugwort leaves, a bunch of forest grass, three lemons, seven to eight bitter willow leaves, a few turkey tail mushrooms, a branch of Mormon tea, 1 teaspoon (1 g) California sagebrush, 3 tablespoons (30 g) dried elderberry, 8 ounces (226 g) brown sugar, and 2 tablespoons (14 g) lerp sugar (insect exudate, or "honeydew"). I boil everything for 20 minutes, cool the solution like any other beer, strain it, and add some wild yeast from a yeast starter made with local juniper berries. I ferment it for 6 days in the carboy, then bottle it (adding ½ teaspoon/2 g sugar to each bottle) and drink it 2 or 3 days later. It tastes a bit like a sour *gueuze* with hints of grapefruit.

LACTO-FERMENTED SODAS

This method of making natural sodas uses whey instead of yeast. When I discovered this method I was extremely skeptical, but it worked like a charm, though it takes a tad longer than using wild yeast.

But it's a good alternative if you're allergic to yeast. If you're lactose-intolerant you might want to try the same method with sauerkraut or similar lacto-fermented juice (I haven't tried that yet).

Making the whey is a cinch. I simply go to my regular grocery store and purchase some plain natural yogurt with live cultures. Back at home, I scoop the yogurt into a strainer, usually lined with cheesecloth (not a must if your strainer has a fine mesh), and let the whey drip into a jar. It's that simple! It usually takes around 6 hours to get the whey from 2 cups (500 ml) of yogurt. Don't waste the yogurt that's left; it's similar to cream cheese.

If you don't use the whey the same day, place the jar in the fridge, where it will keep for at least a week.

6–7 cups (1.4–1.6 L) water

¾–1 cup (150–200 g) organic cane sugar or palm sugar, or ¾–1 cup (180–250 ml) maple syrup or honey

3 lemons, sliced

1 small sprig mint

½ cup (120 ml) whey (from strained yogurt)

Procedure

1. Put the water, sugar, lemons, and mint in a container. Add your whey and cover. You can use any container, or even a jar, but make sure the lid isn't on so tight that fermentation gases can't escape. I place the jar in the fridge overnight for the herbs and lemons to infuse, then strain the contents the next day. You could even keep it a couple of days in the fridge for more flavors.

2. Strain the liquid into clean recycled soda bottles or swing-top glass bottles. Note that you can also add the whey at this stage (after the infusion) if you like.

   The fermentation should take anywhere from 3 days to a week depending on your location and temperature. Here in Southern California, my fermentation was active within a couple of days during the summer. Check the pressure after a couple of days and, if you're happy with it, place the bottles in the fridge. I like to drink it within a week.

An even simpler recipe would be to use ½ gallon (1.89 L) of sugary fruit juice. Stir around ½ cup (120 ml) of whey into the liquid and pour it into recycled soda bottles or swing-top bottles. Check the pressure as explained above and place in the fridge when ready.

There are a lot of yummy fermented concoctions you can make with local fruits or wild berries. There's plenty of inspiration online if you research the subject.

NO-FERMENTATION SODA

While I researched old recipes for "medicinal sodas"—as made by pharmacists in the 19th century—I was amazed to find that a lot of them were actually made using an acid base and baking soda. The principle is that you can add lemon juice, vinegar, or citric acid to water, then pour a bit of baking soda in the container—and voilà! The acid in the water mixes with bicarbonate of soda (baking soda) to create carbon dioxide gas. The generated bubbles of CO_2 make the drink fizzy. If you put it in a closed container such as a bottle, this creates carbonation similar to that found in a fermented soda.

2 cups (500 ml) water

Juice of 2 lemons

2 tablespoons (25 g) sugar, or 2 tablespoons (30 ml) honey

½ teaspoon (3 g) citric acid

½ teaspoon (3 g) baking soda

Aromatic herb(s) of your choice, or flavored syrup, to taste

Procedure

1. Mix the water, lemon juice, sugar, and citric acid.
2. Pour the contents into a 16-ounce (500 ml) recycled plastic bottle or swing-top glass bottle, filling to the bottom of the neck.
3. Using a small funnel, pour in the baking soda and *very quickly* close the lid before the contents gush out of the bottle (it generates a lot of bubbles). You have a couple of seconds.
4. For decoration and flavors, you can always add some tasty herbs inside the bottle and strain the contents while serving. I don't age these sodas for very long and usually drink them within 1 or 2 days of making them. I use this method mostly for fun, as I much prefer making my soda through regular fermentation.

The pressure is pretty mild with this recipe and you can experiment a bit, but please, be careful: The combination of too much citric acid and too much baking soda can generate a lot of pressure. I also advise you to try first using recycled plastic soda bottles (you don't want an exploding glass bottle). Be safe!

Soda infused with California
sagebrush and horehound.

Soda infused with lemonade
berries and yarrow.

Soda infused with hops flowers.

RESOURCES

Plant Identification in the United States

This list includes a few books I'm familiar with, but it's very incomplete. A simple search online or on Amazon.com should help you find plant identification books you can use to learn about your local wild edible plants.

SOUTHWEST

California Foraging: 120 Wild and Flavorful Edibles from Evergreen Huckleberries to Wild Ginger by Judith Lowry

Foraging California: Finding, Identifying, and Preparing Edible Wild Foods in California by Christopher Nyerges

The Forager's Harvest: A Guide to Identifying, Harvesting, and Preparing Wild Edible Plants by Samuel Thayer

Nuts and Berries of California: Tips and Recipes for Gatherers (Nuts and Berries Series) by Christopher Nyerges

NORTHEAST

Edible Wild Plants: A North American Field Guide to Over 200 Natural Foods by Thomas Elias and Peter Dykeman

Northeast Foraging: 120 Wild and Flavorful Edibles from Beach Plums to Wineberries by Leda Meredith

SOUTHEAST

Southeast Foraging: 120 Wild and Flavorful Edibles from Angelica to Wild Plums by Chris Bennett

NORTHWEST

Foraging the Mountain West: Gourmet Edible Plants, Mushrooms, and Meat by Thomas J. Elpel and Kris Reed

Pacific Northwest Foraging: 120 Wild and Flavorful Edibles from Alaska Blueberries to Wild Hazelnuts by Douglas Deur

CENTRAL

A Field Guide to Edible Wild Plants: Eastern/Central North America (Peterson Field Guides) by Lee Allen Peterson and Roger Tory Peterson

Beers, Meads, and Wines

Ale, Beer, and Brewsters in England: Women's Work in a Changing World 1300–1600 by Judith M. Bennett

Make Mead Like a Viking: Traditional Techniques for Brewing Natural, Wild-Fermented, Honey-Based Wines and Beers by Jereme Zimmerman

Radical Brewing: Recipes, Tales and World-Altering Meditations in a Glass by Randy Mosher

Sacred and Herbal Healing Beers: The Secrets of Ancient Fermentation by Stephen Harrod Buhner

Uncorking the Past: The Quest for Wine, Beer, and Other Alcoholic Beverages by Patrick E. McGovern

Wild Fermentation: The Flavor, Nutrition, and Craft of Live-Culture Foods, 2nd edition, by Sandor Ellix Katz

Herbalism

The Herbal Handbook: A User's Guide to Medical Herbalism by David Hoffmann

The Herbal Medicine-Maker's Handbook: A Home Manual by James Green

A Modern Herbal, volumes 1 and 2, by Mrs. M. Grieve

RECIPE INDEX

INDEX

Note: Page numbers in *italics* refer to photos.

wintergreen (*Gaultheria procumbens*)
   in fruit kvass, 227
   in Northeastern Kvass, 225
   in root beer, 131
Winter in the Forest Beer, *150*, 151
wood chips
   in beer, 66–67
   in mountain brews, 149
   in Woodsy Mushroom Beer, 166
woodruff (*Galium odoratum*), 65
wood sorrel (*Cochlearia* spp.),
   68, *247*
Woodsy Mushroom Beer, 166, *167*
woolly bluecurls (*Trichostema lanatum*)
   in chaparral herbal blends,
      124, *125*
   in Hike in the Chaparral
      Soda, *248*
   hot brewing with, 80
   as local ingredient, 72
   in Rural Wilderness Drink,
      252, 253
   in Spring Chaparral Beer, 155
wormwood (*Artemisia absinthium*), 10, 62
wormwood, common. *See* mugwort (*Artemisia vulgaris*)
worts, defined, 33

Y

yarrow (*Achillea millefolium*),
   63, *112*

as base flavor, 148, 149
in beer, 60–61, *75*
in bread kvass, 223
in cactus pear wine, 104
in Chaga Beer, *233*, 236
in chaparral herbal blends,
   124, *125*
cold infusions with, 92
in fruit kvass, 227
in Ginger Beer, 119
hot brewing with, 80
in Late-Spring Mountain
   Beer, 152
in lazy wines, *189*, 191
in meads, 208
in No-Fermentation Soda, *271*
pregnancy precautions, 10
in sodas, *247*
uses of, 113
in wild beer, *52*
in Wild Belgian Beer, 160,
   161, *162*
in Yarrow Beer, 113
Yarrow Beer, 113
yeasts. *See also* wild yeasts
   for cold brewing, 84
   commercial, 36, 40
   defined, 33
   effects on fermentation, 99
   for hot brewing, 81
   role in alcohol content, 21
yellow birch (*Betula alleghaniensis*)
   in beers, 67

in forest beer, *75*
hot brewing with, 80
in Vermont Forest Beer, 159
in wild yeast starter, *32*, *34*
wood chips of, 66
yellow dock. *See* curly dock
   (*Rumex crispus*)
yerba santa (*Eriodictyon californicum*), 70
   as bittering agent, 8
   in fruit kvass, 227
   in Hike in the Chaparral
      Soda, *248*
   hot brewing with, 80
   as local ingredient, 72
   in meads, 202, 203
   in Spring Chaparral Beer, 155
   in Yerba Santa Manzanita
      Beer, 139
Yerba Santa-Manzanita Beer,
   138, 139

Z

Zen of fermentation
   beverages representing an
      environment, 144–45
   turning a hike into mead,
      207–8, 210
   when cold brewing, 104–5
Zimmerman, Jereme,
   200, 201
Zingiber officinale. See ginger
   (*Zingiber officinale*)

ABOUT THE AUTHOR

Pascal Baudar is an author, wild food researcher, traditional food preservation instructor, and wild brewer living in Los Angeles. For the last 17 years, his passion has been to investigate the flavors and possible uses of local wild edibles through extensive research and experimentation in the fields of modern and traditional methods of food preservation, herbalism, and ethnobotany.

A sort of culinary alchemist, he has seen his locally found wild ingredients and unique preserves make their way into the kitchens of such star chefs as Ludo Lefebvre, Josiah Citrin, Ari Taymor, Michael Voltaggio, Chris Jacobson, Niki Nakayama, and many others. Over the years, through his weekly classes and seminars, he has introduced thousands of home cooks, local chefs, and foodies to the flavors offered by their wild terroir.

Pascal has also served as a wild food consultant for several television shows including *MasterChef* and *Top Chef Duels*. He has been featured in numerous TV shows and publications including *Time* magazine, the *Los Angeles Times*, *LA Weekly*, the *New York Times*, and many more.

In 2014 he was named one of the 25 most influential tastemakers in LA and in 2017, one of the seven most creative cooking teachers by *Los Angeles Magazine*.

Pascal offers wildcrafting and traditional food preservation classes through his website: www.urbanoutdoorskills.com.

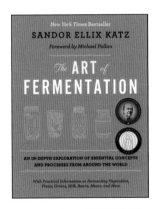

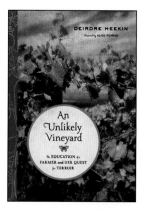
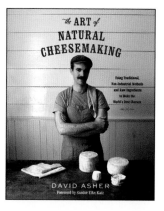